A Forensic Scientist's Guide to Color

Charles A. Steele

DEDICATION

I would never have thought my life would have the opportunities that are before me now, but because of the support and faith of my loving wife Sandra it does. So, let me dedicate this book to you.

CONTENTS

Acknowledgments   i
Forward   ii

Introduction   Pg 1

Chapter 1: Describing Color   Pg 5
1.1 The Science of Color   Pg 7
   1.1.1 The Physics of Color   Pg 7
   1.1.2 Spectrophotometric Color Measurements   Pg 10
   1.1.3 The Biology of Color   Pg 14
1.2 Color Systems   Pg 16
   1.2.1 Historical Color Systems   Pg 16
   1.2.2 Modern Color Systems   Pg 19
1.3 Color Construction   Pg 22
   1.3.1 Color Separation from Background   Pg 26
   1.3.2 Ambient Light   Pg 27
1.4 Implications of Color   Pg 29
Chapter 2: Forensic Applications Using Color   Pg 33
2.1 Blood Spatter   Pg 35
   2.1.2 Coloration   Pg 36
   2.1.3 Chemiluminescence   Pg 36
   2.1.4 Fluorescence   Pg 37
   2.1.5 Color Change   Pg 37
2.2 Visualization of Fingermarks   Pg 38
2.3 Colorant Selection   Pg 40
Chapter 3: Types of Colorants   Pg 44
3.1 Acid Dyes in Forensic Applications   Pg 46
3.2 Basic Dyes in Forensic Applications   Pg 55
3.3 Solvent/Disperse Dyes in Forensic Applications   Pg 62
3.4 Reactive Chemicals in Forensic Applications   Pg 79
3.5 Pigments in Forensic Applications   Pg 83
Chapter 4: Applications of Color Science   Pg 88

Appendix 1: Descriptions of Images   Pg 94
Appendix 2: Glossary   Pg 99

ACKNOWLEDGMENTS

I am of the opinion that meaningful work seldom happens in a vacuum. This work is no exception. Over the length of my career I have had the opportunity to work with and learn from a great many experts in color and forensics. I would list them here, but I am afraid that I would accidentally omit a name. So let me just say to those of you who have helped me learn or worked with me: "thank you."

Forward

When I first met Charles, he was trying to finish his Masters of Science in Forensic Science here at the University of Illinois at Chicago. He had been away from academia for a while and the paperwork involved in completing his task seemed endless. Our first conversation let me know I was dealing with a gentleman interested in advancing his knowledge in new ways to enhance his already successful career and achievements. He completed our program and at the same time infected the students with whom he studied with a greater desire to learn and do well in forensic science. I knew that I wanted to maintain a relationship with him in our program after his graduation.

Charles has a way of passing on information that makes learning effortless and interesting. I thought about information I had lost with time and thought about new ways to pass on information to future students. When Charles let me read this work, A Forensic Scientist's Guide to Color, I did so and remembered much I had forgotten about the elegance of light and the concepts involved in its use in forensic science. I also learned more than I had previously known and did so in a very easy way. I hope you will enjoy this work as much as I did. Whether you are just now learning something of the use of light not only in forensics, but other areas of science, or reading to remember knowledge lost through disuse, you will enjoy the way it is presented and come away more knowledgeable than before.

Albert Karl Larsen, Jr., Ph.D

Director of Graduate Studies in Forensic Sciences &
Director of Graduate Studies in Biopharmaceutical Sciences
University of Illinois at Chicago

Introduction

People are creatures of habit. They tend to do things the way they always have done them. For a variety of reasons however, the "status quo" mentality is a dwindling option for the forensic scientist. Techniques and resources are changing. Advancements within the fields and pressures from external sources are driving changes and improvements in the way evidence is handled and identified.

One need only look at the impact of DNA testing to see how capable modern scientific techniques are of solving crimes that would have been unsolvable even a few years ago. Consider that in 1986, Dr. Alec Jeffreys[1] applied DNA testing to a criminal investigation for the first time. (Gaensslen 2008) Now, slightly less than thirty years later the method has evolved and taken its own place as a distinct discipline in the forensic sciences.

DNA testing has had staggering impact. Not only can forensic labs link arrestees to unsolved crimes committed decades earlier, (Samuels, et al. 2012) DNA analysis can also lead investigators to their suspect through indirect links like familial matches. (Flowers v. State)

---

[1] Subsequently knighted for his work, now Sir Alec Jeffreys.

Of course not every crime can be solved with DNA testing. But the power of this new kid on the block has raised the bar for the expectations of the quality of information generated in the other forensic disciplines. Fortunately, the quality of the analytical tools available to the other branches of the forensic sciences has also improved dramatically in the past few decades.

Some of the various types of spectrum analyzers that once required rooms of their own can now be carried in one hand. Scientists can now take the equipment to the crime scene. These new techniques and equipment are allowing forensic scientists to process more evidence than ever before with a higher degree of precision and accuracy.

However, even with the improvements in technology, or perhaps because of them, there have been some significant errors brought to light. Data from the innocence project blames incorrect scientific work as a contributor in half of the cases of wrongful conviction that they have gotten overturned. In addition, in almost half of the overturned cases, correct scientific evaluations were able to subsequently identify the true perpetrator. (www.innocenceproject.org)

Unfortunately even the power of these new tools cannot be applied in all cases and on all evidence. The financial resources and time to execute simply do not exist. Fortunately not every improvement requires an implementation of "new" science. New techniques can grow from the investigation and application of fundamental principles. This is especially true where the evaluation of physical evidence is based on its appearance or the ability of the analyst to view it.

Consider fingerprint and fingermark comparisons. The analysis of this type of evidence is essentially visual. First the investigator must be able to locate and record the evidence. Then the analyst must be able to discern detail well enough to make a comparison.

As "detection apparatus" the human eye is inconsistent. Humans do not all distinguish shapes and colors equally well. And while to some extent the human eye can be calibrated -

color and forms vision can be tested with color acuity and forms blindness tests - most people do not take these tests and don't really know how well they perceive color and form or what their limitations are. Forensic methods relying on the human eye should therefore be designed to maximize the ability of the viewer discern color and shape. One way to effect this maximization of visibility is with the addition of color.

Forensic scientists have the choice of using a variety of scientifically different techniques that add visualization enhancing colorants to evidence. Very often a technique only has a single colorant option. The Oil Red O process for developing fingerprints on porous samples for example, (Beaudoin 2004) only ever imparts a red color. If the background substrate is red, then the analyst may need to select a different technique. So selection of the proper technique should be based not only on what will work but what will provide the best visualization option. And this brings us to the objectives for this book.

When selecting which color development is best, the analyst should be aware of what shades will show up against the substrate background. The first objective therefore is to provide an explanation of color theory as it applies to forensic science. From this, the proper selection of color addition and lighting can be made.

There are many chemistries and methods available for color addition. All chemistries will not work equally well on all materials. For example: Axis Inversion Dyes work well for providing durable images on plastic materials, but on porous materials like writing paper they are not as useful. The second objective therefore is to explain and categorize the chemistries available for visualization.

The third objective is simply to provide an index of common colorants used in forensic science with specific attention on those used for blood and fingermark detection. There are literally thousands of colorants available. So in selecting the colorants for these lists, an attempt has been made to include the most common.

The final objective of this book is to provide the reader with enough of an understanding of the specific colorants used in the different techniques that they have a basis for expanding the color pallet for a technique they use. Where possible the analyst should separate the technique from the colorant. The technique is the method not the colorant. Under ideal conditions a method would have several color options.

It is entirely possible that the user of this book may find a colorant they use missing. Hopefully, this book will serve as a guide for how to categorizing these colorants so that the same principals can be applied.

References:

Beaudoin, Alexandre; New Technique for Revealing Latent Fingerprints on Wet Porous Surfaces: Oil Red O. Journal of Forensic Identification. 2004:413 54(4)

Flowers v. State, 654 N.E.2d 1124 (Ind. 1995)

Gaensslen, R. E.; Harris, H. A.; Lee, H.; Introduction to Forensic Science & Criminalistics. McGraw, Hill New York. 2008

Samuels, Julie; Davies, Elizabeth; Pope, Dwight and Holand, Ashleigh; Collecting DNA from Arrestees: Implementation Lessons; June 2012, NIJ Journal 270

www.innocenceproject.org/Content/DNA_Exonerations_Nationwide.php

Chapter 1: Describing Color

To the analytical scientist, color does more than just incorporate style elements, it adds valuable functionality. Whether an analyst is looking for a latent fingermark, scanning for blood residue or conducting a microscopic examination, visualization through color differentiation plays a key role. Some evidence is chromatically distinct enough to be seen as it is. But other evidence needs to be enhanced by changing its color or that of the environment around it.

But what is color? What does it mean to say that something is yellow? On the surface this seems like an easy question. But the answer depends on why it's yellow. Does the thing exclusively reflect yellow wavelengths of light and absorb the remaining? Or does the thing actually give off green and red wavelengths of light? Why are both of these phenomena perceived by the human eye as yellow but seen differently by spectrophotometers? The analysis and understanding of color requires the analysis and understanding of physics and human biology.

The definition of color is also affected by its ultimate use. To a painter where specific shades are created by blending paints that absorb light, the primary colors are red, blue and yellow. A proper combination of all three yields black. To optical physicists, where color is a combination of incident light waves, the primary colors are red, blue and green. The proper combination of the three yields white.

So when changing the color of an object to differentiate it from its background, whose understanding of color is best: the painter or the physicist? The answer is that it depends. Both understandings are important in knowing what color to add.

There are those who will give the simplistic advice of: choose a dark color for light backgrounds and a light color for dark backgrounds (Miller 2013). This is sound advice, it has served science well for many years and it is supported by fundamental color theory. A more rigorous approach will however offer the analyst greater resolution capability.

To get the best differentiation between evidence and the background where it resides it is important to know both the necessary direction and method of the needed color change. Therefore, before addressing the specifics of individual colorants and their uses, the analyst should understand the nature of the colors they are working with and the nature of color itself.

Table 1: Color by Wavelength

λ Wavelength (nm) / Perceived color
400 - 440 / Violet
440 - 480 / Blue
480 - 560 / Green
560 – 590 / Yellow
590 – 630 / Orange
630 – 700 / Red

1.1 The Science of Color
1.1.1 The Physics of Color

A discussion of color must begin with the bundles of electromagnetic energy called photons. These bundles travel as waves with specific wavelengths related to their energy[2] and interact with mater and energy in a variety of ways.

Color is the perception of combinations of specific photons that have wavelengths between 400 nm and 700 nm (Knight et al. 2013). Each photon has a specific energy level and is perceived as a specific color as shown in **Table 1**. What a person or machine perceives as the color of an object are the combinations of photons that reflect off, emit from or pass through that object.

Most commonly people see reflected color. Polychromatic light, sunlight for example, strikes a surface, a green leaf for example (**Image 1**). Then some portion of the incident light is reflected and viewed.

In the leaf example, the chlorophyll and natural pigments absorb some of the incident photons with specific energies and reflects others. So although the incident light is a combination of photons having all the visible wavelengths and is perceived as white light, the reflected light is combination of fewer photons of only particular wavelengths and is viewed by the human eye as green.

Image 1: Color of a leaf

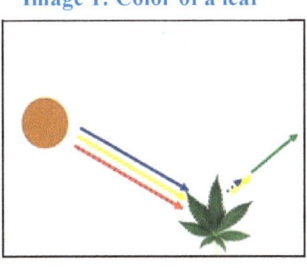

---

[2] The relationship is given by: $eV = \hbar c/\lambda$ where: eV is the energy of the photon in electron volts, $\hbar$ is Planck's constant, c is the speed of light and λ is the wavelength.

When light passes through an object more photon/matter interactions need to be considered. First, as with reflection, some photons are absorbed by the sample. Other photons pass through freely. Finally, some photons are reflected (scattered) back the way they came. The net effect is that transmitted color may be different than reflected color because of the difference of these scattered photons in addition to the loss of the absorbed photons.

A transparent or translucent object may therefore have more than one color depending on where the observer is relative to the light source. Take for example a semi-transparent sheet (**Image 2**), which absorbs some photons of particular wavelengths, reflects some of different wavelengths and allows the remaining to pass through. An observer seeing the reflected light would see the sheet as being one color and a person seeing the transmitted light would see the sheet as being another color because each is seeing a different set of photons. The specific shade that is perceived by the human eye is the net combinations of all of the visible light photons that reach it.

Image 2: Color of Semi-Transparent Objects

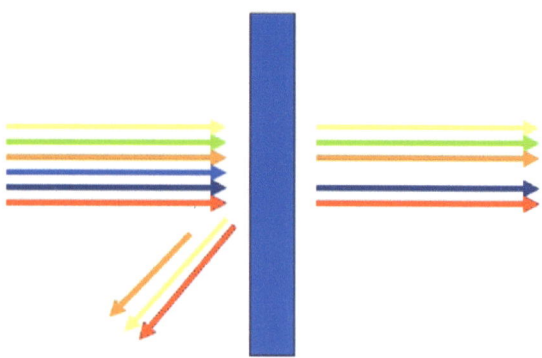

The third color effect that should be considered in the context of forensic science is emitted color. This category includes: Fluorescent, Phosphorescent and Luminescent effects.

Surface electrons in light emitting compounds are excited by a variety of means (heat, chemical reaction, incident photons, etc.) resulting in the production of photons.

With fluorescent compounds, an electron is excited by an incident photon which transfers some or all of its energy to the electron via collision. The now more energized electron moves into a higher orbital positon know as a meta-stable state. This higher orbital is not a natural state for the electron and is unstable. The electron will soon fall or decay back to its natural position known as its rest state. As the electron decays to its rest state it gives off one or more photons equal to the energy it absorbed.

As a practical matter there is energy lost in the process. The energy of the photons given off by the decaying electron will be less that the energy of the incident electron. The emitted photon is therefore of a higher wavelength than the incident photon. Thus invisible UV light can strike Rhodamine and produce a glow in the visible range. This effect is rapid enough that when the incident light is removed the glow ceases.

Phosphorescent compounds are similar to fluorescent molecules. An incident photon strikes and excites a surface electron. However, the now excited energy of the electron is shared with other bonded atoms. Because of this sharing the energy dissipates more slowly and the produced glow can persist for a long time after the incident light is removed.

Luminescent effects are triggered by energy from sources like heat and chemical reactions. But as with the previous two examples, the excited electron decays giving off light.

Regardless of the source, perceived color is the net effect incident photons which can be seen by the eye or scanned with an instrument.

1.1.2 Spectrophotometric Color Measurements

Colorimeters and Visible-Light Spectrophotometers analyze light one wavelength at a time and generate a spectrum in the full visible range. Therefore they have the ability to distinguish two samples, even if they are the same color to the human eye, if they have different composition.

Colorimeters and spectrophotometers typically shine light from a source through a sample or reflect light off of a sample. The light is then collected and measured by some sort of detector. For the purpose of discussion consider **Absorption Spectroscopy**. An absorption spectra is collected by shining light through a sample and measuring the Transmittance (T) at specific wavelengths (λ).

Transmittance is the ratio of the light that passes through something relative the incident light.

$$\mathbf{T} \quad = \quad \mathbf{I/I_0} \qquad\qquad (1)$$

Where:

$\mathbf{I}$ = The radiation intensity reaching the detector through the sample.

$\mathbf{I_0}$ = The radiation intensity reaching the detector without a sample in the optical path.

$\mathbf{T}$ describes a non-linear curve and is therefore frequently translated to Absorption (**A**) at a specific wavelength (λ) according to equation (2).

$$\mathbf{A}_\lambda = \mathbf{-logT = log\ I_0/I} \qquad\qquad (2)$$

The Beer-Lambert Law equates this **A** to properties of a sample according to equation 3 below. (Roberts 1885)

$$A_\lambda = \varepsilon lC \tag{3}$$

Where:

ε = Absorptivity, usually referred to as the Extinction Coefficient. ε is a property of the sample being scanned.

l   = Path Length. This is the distance that the incident radiation covers while passing through the sample.

C   = the molar concentration of the sample being scanned.

If multiple absorbing species are present (3) can simply be expanded to:

$$A = \varepsilon_1 lC_1 + \varepsilon_2 lC_2 + \varepsilon_3 lC_3 \ldots \tag{4}$$

In the case of reflection spectra, the photons are bounced off of a sample, as mentioned before, some are absorbed and some are reflected. The reflected photons are directed to a detector which, similar to absorption spectroscopy, records the amount of light.

Unlike absorption spectroscopy the materials do not simply absorb photons or allow them to pass through. The path of the reflected photons are bent according to the refractive index (n) of the material.[3]

The absorbed photons are still calculated by the Beer-Lambert formula, but this is compiled with the bent photons. Putting it altogether the Fresnel equation (5) describes the intensity of the reflected photons. (Hecht 2002)

---

[3] When measuring reflected spectra, the index of refraction is dependent on the extinction coefficient.

$$R = [(\boldsymbol{n} - 1)^2 + K^2 \,] \,/\, [(\boldsymbol{n} + 1)^2 + K^2] \qquad\qquad (5)$$

Where:

R = Reflection measurement

$\boldsymbol{n}$ = The complex index of refraction **(n + iK)**

n = Normal index of refraction

K = The absorption index

In either spectroscopic method, the absorption of the photons by the material is an important factor. The validity of this relationship allows the determination of the concentration of a component that absorbs a specific wavelength when $\mathcal{E}$ is known. However, three things must be considered when applying this relationship.

Consider first that, $A = \mathcal{E}lC$ describes an observed relationship of photons being absorbed by matter and that this relationship is only valid and linear for specific concentrations of photons and the absorbing matter. In addition to this balance, the sample must also be dilute enough that the average distance between the absorbing molecules is too great to allow one molecule to affect its neighbor's charge distribution. Typically therefore samples should be less than 0.01M (Skoog 1985).

The second important issue to consider is that $\mathcal{E}$ is not a physical property or a unit like voltage or gram. It is an interpretation of the decrease of the sum of the photon energy impacting a detector which is neither 100% efficient, nor necessarily linear in its own response from 0 eV incident energy to saturation. This means that different detectors may respond differently to the same photon concentrations and wavelengths.

Each photon incident on the detector has an energy in eV described by the following equations. (Sears 1987)

$$eV = \hbar c / \lambda \tag{6}$$

Where:

$\hbar$ = Planks constant.
c = The speed of light.
λ = The photon wavelength in nm.

Using equation (6), 500 nm photons are 120% as energetic as 600 nm photons. A photomultiplier detector that responds to photons over this whole range and is non-linear near n 600 nm photons would be non-liner to n/1.20 500 nm photons.

More importantly, several types of detectors exist. Diode Arrays for example, have multiple detectors each responding to smaller wavelength ranges. Each type of detector has a different response to the light that strikes it. As a result, scans of the same sample on different machines can yield different absorption readings. It should also be noted that the efficiency and response of detectors decline over their life of use.

Fortunately, for a spectrophotometer in good working order there will typically be a linear range where A/C remains reasonably consistent. Determinations of $\mathcal{E}$ can be done in this range.

This introduces the third consideration. When using the Beer-Lambert law in the linear region of an absorption curve, A = $\mathcal{E}$lC describes a line. Any line in a 2-dimensional (x,y) plane can be described by the equation y = mx + b; where m is the slope and b is the y intercept. $\mathcal{E}$lC must therefore equal some definition of mx + b.

The intercept, b is determined by the energy level of the detector at the selected "zero" point. The slope, m is dependent on the quantum response of the detector itself. Therefore, even for the same sample, the slope and intercept of the line can be different from one machine to another based on the functionality of the detector.

All comparisons of $\mathbf{A}_\lambda$ should therefore be done on a single piece of equipment, or machines that have been calibrated to each other.

When all of the above is properly accounted for, visible spectroscopy can provide extremely good data on the concentration or identity of a compound. However, one must be cognizant that compounds and blends can mimic the spectra of other compounds and blends. Therefore, except in special cases, visible spectra should typically only be used to corroborate identity or determine concentration.

Perhaps a more significant limitation is that most spectrophotometric equipment isn't designed to locate evidence, only analyze it once it has been collected. Visual color acuity therefore is therefore also critical for the forensics sciences.

1.1.3 The Biology of Color

The photon receptors in the retina of human eye are called: Rods and Cones. A human eye can have more than 100 million rods which are highly receptive to the energy of the photons impacting them. Rods are sensitive to intensity, but do not distinguish specific photons or by extrapolation, color.

The retina also has 6 to 7 million cones distributed among three[4] types: long wavelength receptors (L), medium wavelength receptors (M) and short wavelength receptors (S). (Schacter 2011) Each type of cone is receptive to specific, slightly overlapping ranges of photons. Color and forms acuity comes from the cones and there are two competing schools of thought on how the cones perceive color: Opponent Process Theory and Trichromatic Theory. (Hurvich 1981)

---

[4] Social networking and internet sources pose the possibility of humans called tetrachromats with 4 types of cones. As of the writing of this book the prevailing science indicates that if such mutations exist they will be exceedingly rare.

Opponent Process Theory suggests an antagonistic system where the overlapping ranges of sensitivity allow for a comparison of the differences between the detected photons along opponent channels: red to green, blue to yellow and black to white. One aspect to this theory is that opposite opponent colors are never perceived together; i.e., there is no red shade of green.

Trichromatic Theory,[5] is based on the discovery that the cones have different absorptions specific to different wavelengths. Each cone then is believed to directly detect a specific color or set of colors. How well each cone can detect its specific color is dependent on the amount of the proper amino acids present.

Each theory leads to different systems for describing color which will be discussed in the next section. What is important to know for this discussion is that regardless of which theory is correct the specific color and forms acuity of a person depends on the number, distribution and overall make-up of the rods and cones in their eyes.

Different people have different distributions of rods and cones. This means that different people also have differing perceptions of colors and forms. Therefore understanding how to adjust the observed colors to improve the ability of the observer provides the maximum ability for anyone to find and analyze the evidence.

---

[5] Also known as the Young-Helmholtz theory of color vision.

1.2   Color Systems

Regardless of the source of the color and the analysis method, color is usually described in terms of the net effect. Many systems have been developed to describe this effect in terms that can be verbally or numerically related to allow for easy exchange of color information. Some of these systems are based on the Opponent Process Theory and some are based on Trichromatic Theory.

1.2.1 Historical Color systems

There have been many systems developed to describe color. The ancient Islamic tradition for example, described color by combining two aspects. The first aspect was a three color triangle bounded by white, black and sandalwood. The second aspect was a four color matrix bounded by red, yellow, green and blue. These two aspects were combined to form a 28 element cycle.

Mystical and religious significance aside, in the Islamic tradition we see the beginnings of the modern understanding of color by breaking color space into aspects. It is in this system that we also first see the separation of light and dark from the other color axes. (Islamic Tradition www.colorsystem.com)

By the late pre-Enlightenment the astronomer Fr. Aron Sigfrid Forsius published what may be the first treatise on color and described a system.[6] The Forsius system was based on what he called "elementary" colors: red, yellow, green and blue mediated by and generated from white and black. The system can be thought of as foreshadowing the opposing color pairs proposed by Ewald Hering centuries later.[7] (Rosen 2010)

---

[6] A.S. Forsius proposed his system in 1611.
[7] Ewald Hering proposed opponent color theory in 1892

Image 3: The Forsius Color System

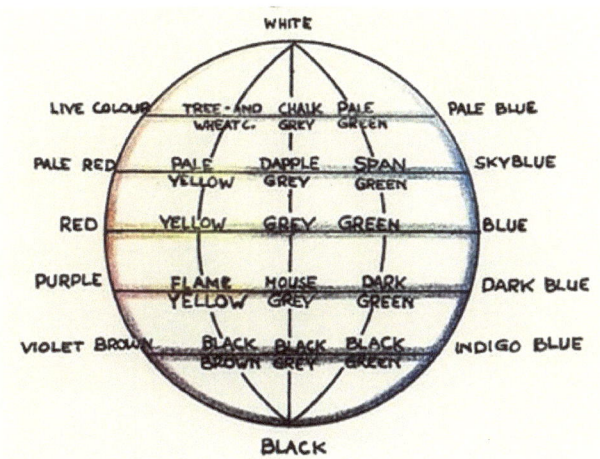

The Forsius system has similarities to modern color systems and in practical terms opened the door for color communication and systematic study. Subsequently, many notable artists, scientists and philosophers have helped further evolve color definition.

In 1704 Sir Isaac Newton diagramed a system based on additive synthesis. What is significant about Newton's systems is that his colors are built around a white center which introduces the aspect of color intensity. The further from the white center the darker the color was.

The aspect of intensity, although new to the understanding of color, is similar to the Forsius system's acknowledgment of a relationship between white, black and color. Also similar to the Forsius system is the oppositional relationship between colors. The Newton color wheel however, evolves both of these concepts into a form more closely in line with modern understanding.

Image 4: Newton Color Wheel

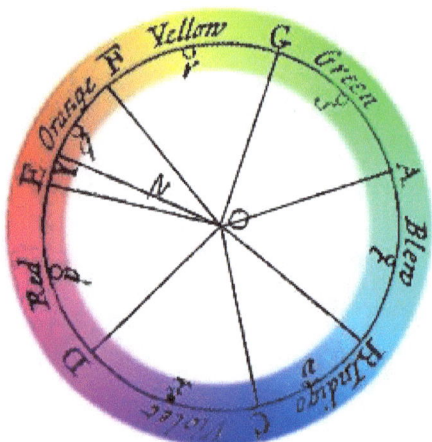

In the Newton system, colors blend with each other in a way that foreshadows the primary shades used today. The color intensity grows as the "whiteness" fades. Perhaps most significantly, in the Newton color wheel we also see that blue is moving closer to being opposed to yellow and green is moving toward being opposed to red.

By the late 1800's Hering's natural color system offered a connection between the color and how the eye perceived it. By the 1900's Munsell and others offered a variety of improvements until we get to the modern systems.

1.2.2 Modern Color Systems

The modern view of color space begins with the ocular response experiments of the 1920's and 1930's by researchers like William David Wright and John Guild. (Wright 1928, Ivaluehue 2010). These experiments quantified the response of the human eye to different color and intensity.

In 1931, the Comission International de L'Eclarige (CIE) took this data and produced a system for specifying color characteristics on a coordinate system based on Trichromatic Theory. These tristimulus axes, labeled x, y and z, accounted for luminance, blueness and the remaining color space.

The color values generated by this system were analogous to the way that the human eye perceived color. The axes proposes by the CIE incorporated both the position in "color space" as well as how the human eye perceives the color. There have been improvements and variations on the system over the years, but it is still the basis for color communications in many industries and even in spectrophotometric testing.

RGB (Red, Green, and Blue) systems, arguably some of the most common systems encountered, also have their roots in these early experiments. Like the XYZ systems, RGB systems follow Trichromatic Theory. These systems describe color in terms of additive combinations of the three primary colors common in physics. They are still typically used for projection systems, television, computer monitors and the like.

In 1948 The Hunter Lab system offered an explanation of color space based on Opponent Process Theory. In the Hunter Lab system the axes became L, a and b (Lab). The critical aspect of this system is that it establishes three perpendicular axes: White to Black (Lightness), Blue to Yellow (b) and Red to Green (a).

Image 5: LAB Color Axes

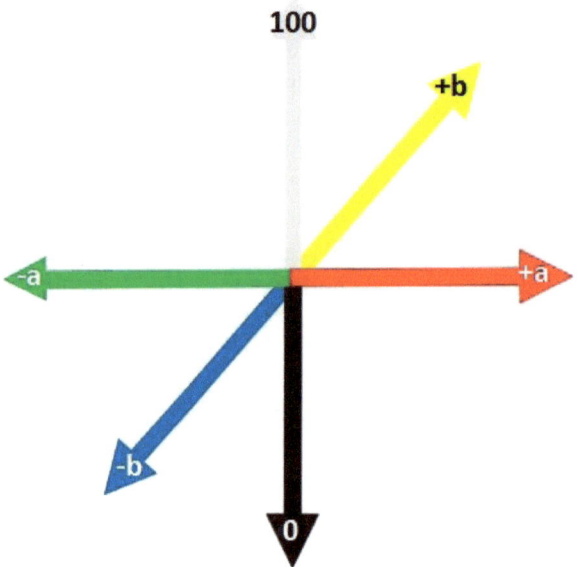

These perpendicular axes represent a critical evolution in the understanding of the perception of color as it recognized that changes in one axis do not affect position on another. For example, making something bluer will not make it less red. So when trying to make things blend or differentiate, it is important to adjust along the correct axis. The Hunter Lab system is still used in many industries.

In 1976 the CIELab system was introduced further refining the view of color space. The primary refinement in the CIELab system is in the math that describes each axis. With both the Hunter Lab and the CIELab systems, the axes are not linear. The Hunter Lab system defines position following a square root of the XYZ coordinate system and the CIELab system follows a cubed root of the XYZ coordinate system. In addition, in the Hunter system the axis expands in the blue region and in CIELab it expands in the yellow.

The CIELab system describes a color space that closely emulates how the eye perceives color. In this system and others like it, see we see that two identically appearing colors can be differentiated by doing something as simple as adjusting the intensity of one or the other.

The CIE color system also gives a method of distinguishing color difference, DE, as the difference between two points in the three dimensional color space. Essentially this is the hypotenuse of triangle made from the hypotenuse of the triangle made from the difference on the "a" axis with the difference on the "b" axis and the

Color acuity varies from person to person but on average people see differences of DE > 1.0

triangle made from the difference on the "b" axis and the difference on the "L" axis. (Heidelberg 1999)

Although color vision varies person to person, industrial applications set their specifications following the assumption that the average person does not perceive color differences less than DE 1 to 1.5 (Steele 1999).

The last modern systems that will be presented here are the CMYK (cyan, magenta, yellow, black[8]) systems. These systems, also called Process Color, are a subtractive method of building shade based on blending the painter's view of primary

---

[8] K is used for black because B was already used for Blue.

colors and darkening with black. CMYK systems are commonly used for printing applications.

The term subtractive color comes from the fact that materials like paint and ink absorb some wavelengths of light. White light initially has all available wavelengths present. If something causes wavelengths to be filtered out, like the film example in section 1.1.1, the remaining wavelengths blend into a color other than white. Which color depends on which wavelengths remain.

The same thing is true of reflected light. When white light reflects off of a surface, if that surface absorbs some of the wavelengths, the reflected color will be something other than white. Once again, what color depends on which wavelengths remain.

Each paint, ink or other blendable material has a specific color determined by which wavelengths it absorbs. When multiple materials are blended, the blend absorbs all of the wavelengths that each of its components individually absorbs. Therefore with each added component, additional wavelengths are subtracted from the white light.

1.3   Color Construction:

As mentioned previously, the "color" of a photon is related to its energy. Different objects absorb and reflect different combinations of photons. It doesn't matter what the object is, if it reflects or emits a 540 nm photon, the photon has the same energy and color as a 540 nm photon from any other source. In addition, the "perception" of color is the simultaneous viewing of combinations of photons. Perceived color is therefore Additive even if the system that generated the shade being view is Subtractive.

Photons are waves and like any other wave, they will constructively and destructively interact adding or subtracting amplitude.

Image 6: Wave Addition

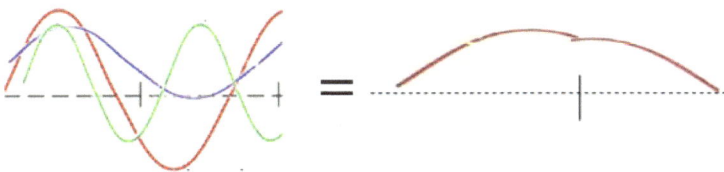

This also means that several different combinations of photons can potentially yield the same net combination and thus produce the same color. Take for example the sample absorption curves below. Image 7 shows three different dyes: a blue, a red and a yellow. As can be seen in the image, all three have a distinctly different absorption (**A**) vs. wavelength (λ) curves. All three absorb different amounts of the available photons between 400 nm and 800 nm.

Image 7: A/ λ Curves of Three Dyes

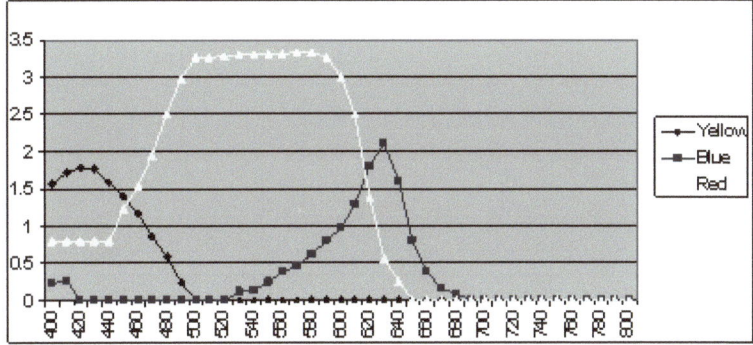

The next curve is a monochromatic brown dye, meaning it has one color molecule.

Image 8: Monochromatic Acid Brown Dye

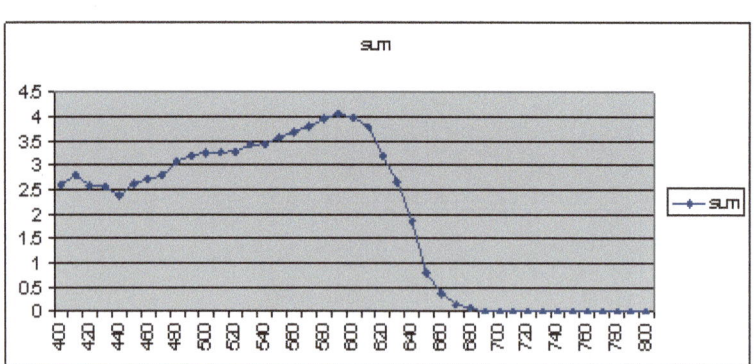

By combining proper amounts of the red, yellow and blue from Image 7 we can produce an absorption graph equal to the brown dye in Image 8.

Image 9: Curve Addition of Dyes in Image 7

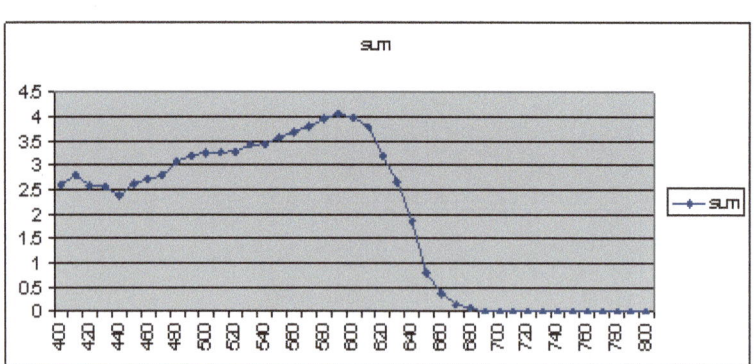

The color produced by the proper combination of the three dyes is visually identical to the single brown dye.

In most cases however the color curves, and by extension the reflected photons, are not so exact. In the example below assume that Brown 2 and Acid Brown are both visually similar colors with a DE of 0.80. The average person would not be able to tell them apart.

Image 10: Brown Dye Curves

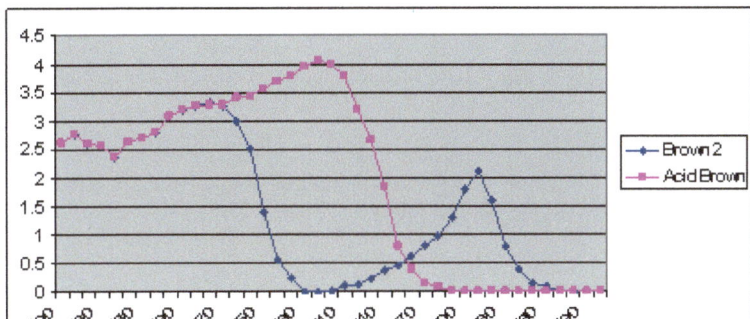

This is the situation most commonly encountered in perceiving colored evidence from a similarly colored background. Either the colors are similar enough that the evidence doesn't stand out from the background; or the sample is too faint and the color of the background overwhelms the visualization.

1.3.1 Color Separation from Background

In forensics science, as in other analytical sciences, the techniques and methods used are designed to allow the examiner to differentiate the color of the sample from the color of the background. The methods of differentiation vary. For example, one common approach to separate the color of the sample from the color of the background is to use a narrow band-pass filter.[9]

In the example displayed in image 10 above, if the Acid Brown was the background and Brown 2 was the evidence and a 515 nm band pass filter were used, the evidence would not be visible. The reflected light at 515 nm is the same for the sample and the background. By contrast if a 600 nm band pass filter were used the evidence would clearly stand out from the background.

Similarly, if we adjusted the available light so that only 460 nm light was available for viewing we would not see the evidence against the background; whereas a 600 nm light would show the evidence clearly. But if we use a 650 nm light, then the difference is only in intensity of the reflected light and the evidence may or may not be missed.

To further enhance the differentiation against the background, colorants can be added to the sample or the background to change enough of the relative color curves of one or the other so that a difference can be easily perceived.

---

[9] In optical uses band-pass filters are colored lenses that block certain wavelengths of light. They are usually produced to only allow narrow ranges of light to pass through. However, any colored transparent material can be used if it blocks the proper wavelengths. Photographer's lighting filters and gels can often be more economical than laboratory band-pass filters and can be just as effective.

1.3.2 Ambient Light

In the descriptions of color theory presented herein, it is assumed that ambient light is "white light" and that wherever you are the sum of the available photons is consistent. However, this is not the case. In fact, ambient color is more or less unique to location.

People spend much of their lives indoors. The color of interior lighting is initially determined by the selection of light sources. Fluorescent bulbs do not emit the same wavelengths as incandescent bulbs or candles. The light from the light sources is then deliberately altered by lamp shades, lenses and other filters placed over the source. Finally the interior light is also reflected off of the colored materials in the area.

Wall paint, floor coverings and even objects in a room all reflect a specific color of light that will mix with the light from the source to give the net illumination. This means that the perceived color in any part of a room will depend of where the observer is relative to the source and any reflective surfaces. As might be expected, this also means that the perceived color of ambient light is not going to be identical in all parts of the same room.

Even outdoors under the sun, ambient light is not uniform. Similar to light in a spectrophotometer, as sunlight passes through the atmosphere some percentage of incident photons are absorbed and reflected. The amount of atmosphere that light passes through is directly related to the amount of photons absorbed. A longer path through the atmosphere means that more photons are absorbed and a greater shift in the color of ambient light occurs. Therefore the intensity of sunlight varies based on location.

For example: one study measured the total annual UV radiation in Chicago, IL as 51,005 MJ/m^2 and 70,005 MJ/m^2 in Prescott AZ. As the earth changes attitude and position in its orbit, the angle and distance that sunlight passes though the atmosphere changes. So even at set locations the available illumination will change day by day.

Time of day will also affect make up of ambient light. Steele and Waring measured the intensity of sunlight at 340 nm, in Chicago at 30-minute intervals for and 8-hour period. The average intensity was 0.240 W/m^2 with significant variation. The peak period was between 10 a.m. to 2 p.m. During this period the intensity varied 33% from 0.200 W/m^2 to 0.298 W/m^2. (Steele and Waring 2003)

It is not just the intensity of the light that varies, composition of ambient sunlight also varies with the composition of the atmosphere itself. Heavy cloud cover can block most of the sunlight leaving ambient light weak and dull. Atmospheric pollution will act as a filter. In extreme cases, this can make ambient light orange to brown.

And of course, just as light reflects off of interior surfaces indoors, sunlight reflects off of objects and surfaces outdoors.

To understand just how significant these variations can be, consider light fading experiments. Light fading is the result of the photochemical reaction between specific incident photons and the electrons in colorant molecules. The differences in structure and chemistry of different colorants make them susceptible to different specific photons. (McQuarrie 1983) By looking at how quickly different colorants fade in different locations one can gauge how similar or different the composition photons are in those locations.

The Steele/Waring experiment exposed three different colorant samples (A, B and C) to equivalent illumination, based on intensity and duration in multiple locations and determined that the relative light fading of the three samples changed location to location. For example, in one group sample A had the greater fade in Monterrey Mexico than Chicago, IL., while in another group, sample B faded more in Chicago, than in Monterrey. (Steele and Waring 2003)

1.4 Implications of Color

The forensic scientist is going to encounter evidence of multiple colors on backgrounds of multiple colors in the presence of a variety of colors of ambient light. Therefore a "one size fits all" approach to evidence visualization is not going to produce the best results in all cases. Fortunately several options exist.

As was mentioned earlier, one approach is to use light sources that emit specific wavelengths of light. Another approach is to view evidence through filters that only pass specific wavelengths of light. These approaches are scientifically sound, but often require expensive equipment and can require the examination of evidence multiple times to use different wavelengths.

A third option is to apply colorants to the evidence being examined. Many colorants exist. Some of the more popular ones will be discussed in the next chapter.

Key Points

- The color of an object is the perception by an eye or artificial detector of the photons which emit from it, pass through it or reflect off of it.

- There are many systems for describing color.

- Scientists tend to describe color in Additive systems based on Red, Blue and Green. Artists tend to describe color in Subtractive systems based on Red, Blue and Yellow.

- Modern systems like CIELab, use a three dimensional color space based on the perceptions of the human eye. In these systems alterations along one axis do not affect the perception along the remaining two.

- People and instruments do not all perceive the same color the same way.

- Most People perceive color differences (DE) of greater than or equal to 1 unit on the CIELab system.

- Ambient light differs location to location.

- Color perception can be modified by altering: ambient light, adding a filter or adding a colorant.

References:

Guild, J. "The colorimetric properties of the spectrum". *Philosophical Transactions of the Royal Society of London. Series A, Containing Papers of a Mathematical or Physical Character.* (1932)

Hecht, Eugene. *Optics* (4th Ed.). Addison Wesley. (2002)

Heidelberg CPS, CIELAB Color Space, Heidelberg CPS, Americas 1999

Hurvich, Leo; *Color Vision.* Sinauer Associates (1981)

http://ivaluehue.blogspot.com/2010/09/pre-newtonian-color-systems.html 2010, Rosen, Dan. *Color Theory: Pre-Newtonian Color Systems (Up to Forsius)*

Knight R., Jones B., Field D., College Physics, a Strategic Approach 2nd ed. Pearson Education Inc. 2013

McQuarrie, Donald, Quantum Chemistry, University Science Books, 1983, pp 12-15.

Miller, Allen Choosing the Best Powder for your Scene, Evidence Technology Magazine September-October 2013, 20-21.

Roberts, R. et.al., Modern Experimental Organic Chemistry 4th Ed. Holt Rinehart and Winston, Inc. 1885

Rosen, Dan. *Color Theory: Pre-Newtonian Color Systems (Up to Forsius)* retrieved from http://ivaluehue.blogspot.com /2010/09/pre-newtonian-color-systems.html 2010

Schacter, Daniel L.; *Psychology Second Edition.* Worth Publishers, New York (2011)

Sears F., Zemansky M., Young D. University Physics 7th ed. Addison-Wesley Publishing. 1987

Skoog D. Principles of Instrumental Analysis 3rd ed. CBS College Publishing, 1985

Steele, Charles; Waring, Christopher; Locational Variations as an Obstacle to Single Point Reference Light Fade Studies AATCC Review April 2003

Steele, Charles; Westman, Mort. Keystone Quality Hairdyes: Technical Guide and Formulary 1st Edition, Keystone Aniline Corporation (1999)

Wright, William David (1928). "A re-determination of the trichromatic coefficients of the spectral colours". *Transactions of the Optical Society* 30 (4): 141–164

www.colorsystem.com/Islamic Tradition

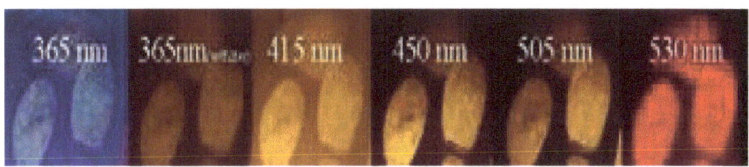

Chapter 2: Forensic Applications Using Color

Forensics Scientists use a variety of colorants to aid them in the finding and analyzing of evidence. The table below lists colorants used for Blood, Fingermark and other applications.

Table 2: Colorants Used in Forensic Applications

Common Name	Type[10]	Class	Use	Color
ABTS	R	Reactive	Blood	Green
Acid Fuchsin (Hungarian Red)	D, F	Acid Dye	Blood	Violet (FL Yellow)
Acid Yellow 7	D, F	Acid Dye	Blood, Fingerprinting	Fl Yellow
Acid Yellow 17	D, F	Acid Dye	Blood, Fingerprinting	Fl Yellow
Amido Black 10B	D	Acid Dye	Blood	Blue/Black
Basic Yellow 40	D F	Basic Dye	Blood	Fl Yellow/Green
Basic Red 14	D F	Basic Dye	CA fumed Fingerprints	Fl Yellow/Orange
Basic Red 28	D F	Basic Dye	CA fumed Fingerprints	Fl Yellow/Orange

[10] D = Direct Colorant, F = Fluorescent, L = Luminescent, P = Pigment, R= Reactive, S = Sublimation

Table 2: Cont.

Common Name	Type[11]	Class	Use	or
Blue Star	R, L	Reactive	Blood	Luminescent Blue/White
Blueblood Tracker	R,D	Acid Dye	Blood	Blue
Carbon Black	D	Pigment	Fingerprinting	Black
Crocein Scarlet 7B	D	Acid Dye		Red to Dark Purple
DFO	R	Reactive	Blood	FL Yellow
Diaminobenzidine	R	Reactive	Blood	Brown
Disperse Blue 60	S	Disperse dye	Fingerprinting	Blue
Disperse Red 60	S	Disperse dye	Fingerprinting	Red
Disperse Yellow 211	S	Disperse dye	Fingerprinting	Yellow
Fluorescein	R, F	Solvent/Acid Dye	Blood	Fl Yellow
Iodine	R	Reactive	Fingerprinting	Yellow to Orange
Leuco Crystal Violet	D, R	Basic Dye	Blood	Red/Violet
Leuco Malachite Greene	D, R	Basic Dye	Blood	Green
Luminol	R, L	Reactive	Blood	Luminescent Blue/White
Nile Red	D, F	Solvent Dye		Fl
Ninhydrin	R	Reactive	Blood, Fingerprinting	Pink
Oil Red O	D	Solvent Dye	Fingerprinting (Porous)	Blue/red
Phenolphthalein	R	pH indicator	Blood	Pink
Rhodamine 6G	D, R	Basic Dye	Fingerprinting	Red/Orange & FL Yellow
Silver Nitrate	R	Reactive	Fingerprinting	Brown
SY43 Solvent Yellow 43	D, F	Solvent Dye	Fingerprinting, Tool Mark	Fl Yellow
Ardrox	D, F	Penetrant	Fingerprinting (CA Fuming), Tool Mark	Fl Yellow
CN-Yellow	D, F, S	Polymer	Fingerprinting	Fl Yellow
Fuming Orange	D, F, S	Polymer	Fingerprinting	Fl Yellow
Metal Check Fluid	D, F	Penetrant	Fingerprinting, Tool Mark	Fl Yellow

---

[11] D = Direct Colorant, F = Fluorescent, L = Luminescent, P = Pigment, R= Reactive, S = Sublimation

While most forensic fields have some use of color, the disciplines of blood spatter analysts and latent fingerprint analysis heavily rely on the application and use of colorants. Often times each colorant is linked to a specific technique, for a specific evidence type. Therefore before discussing the chemistries physical properties of specific colorants, a quick explanation of these evidence types and the use of color with them may be helpful.

2.1 Blood Spatter

Bloodstains can be found at crime scenes or on physical evidence. The shape of these stains may be important and even if it is not, the stain still contain a number of components useful to the forensic scientist: proteins, amino-acids, plasma and of course blood cells (Red, White and Platelets), (Bremmer, et al. 2011). These components can be used to reconstruct events or to perform identification through traditional markers or DNA analysis.

Red blood cells comprise up to 97% of blood content. They are non-nucleated and thus do not contain DNA. However red blood cells do contain hemoglobin, carrying oxygen and iron and are therefore important to the chemical reactions used in blood detection.

There are many fewer white blood cells than red blood cells (1:700). White blood cells are nucleated and contain the necessary the DNA for genetic identification.

Blood plasma has serum proteins which are also useful for DNA testing as well as albumin and immunoglobulins useful for many traditional tests (Meyers, et al. 2006).

The finding of blood stains is based on the ability of the investigator to distinguish it from the background. In some cases this is easy. In some cases it is not. When an investigator gets to a crime scene the blood is usually: dried droplets, large patches or faint residue below the visible threshold (Bremmer, et al.

2011). It is the faint residue that needs the addition of color to aid detection.

Blood residue may be hard to detect due to concentration, attempt at removal or simple erosion due to time and weather. Fortunately several techniques exist for this task including: visual inspection, alternate light sources and chemical treatment.

Many of the chemical treatments react with the hemoglobin in the blood and produce a visible reaction that may be optically or physically recorded. It should be noted that these detection chemicals can react with more than just blood, so their use is limited to presumptive tests for the presence of blood.

2.1.2 Coloration

Many chemicals, pH indicators and dyes will directly color blood. Two common examples are Amido Black and Phenolphthalein (discussed below). Amido Black, is a common acid dye used to physically color the blood. Phenolphthalein is an acid-base indicator that transitions from colorless to red/pink in contact with blood (Grodsky & Kirk 1951, Cox 1991).

2.1.3 Chemiluminescence

On its own, blood does not fluoresce or emit light, however, fluorescent reactions are often used with blood detection Bluestar® Forensic, and Luminol for example will both emit a glow in the presence of blood. (Barni et al. 2007, Sahs 1992). The reaction is based on the iron present in hemoglobin and catalyzed with peroxide. The intensity, wavelength and duration of the emitted light is related to the reaction rate and efficiency as well as the concentration of hemoglobin. (Tontarski et al. 2009)

These reactions are good for finding blood stains in dark areas but are short duration (Barni, et al. 2007, Cox 1991) and often require reapplication which may dilute DNA or other biologic markers. They are however useful to help find blood stains below the visible threshold, even in cases where an attempt has been made to remove the stain (Laux 1991, Cox 1991).

2.1.4 Fluorescence

Similar to chemiluminescence, reactions with chemicals like Fluorescein and Diaminobenzidine emit light when exposed to and excited by the proper wavelengths of light. A reduced form of fluorescein, for example, is reacted with hemoglobin in the presence of peroxide and oxidized to fluorescein. When this dye is excited with wavelengths between 380 - 485 nm, an emitted light is produced. This is often viewed with a filter to enhance visualization (Cheeseman 1999, Cheeseman & DiMeo 1995, Cheeseman & Tomboc 2001). This reaction is good for finding traces of blood below the visible threshold, especially in dark places. It should however be noted that the reaction is exothermic and can be a fire risk.

2.1.5 Color Change

Leuco Crystal Violet (LCV), Leuco Malachite Green and Blueblood™ Enhanced are completely reduced and virtually colorless forms of common dyes. (Bodziak 1996). Like other leuco dyes, the color change is permanent and strongly visible against a variety of backgrounds.

2.2 Visualization of Fingermarks

Simply put, the Locard[12] principal of cross-transfer states that when people interact with a place, they take some of the place with them and leave some of themselves behind (Saferstein 1990). Among the most useful pieces of transfer evidence are fingerprints.

Fingerprints are the images left when the friction ridges of finger tips are pressed into a suitable medium. They have been used since ancient times, as far back as Babylon and ancient Egypt as a means of identification. (Laufer 1912). From a forensic perspective, fingerprint analysis starts in the 1820's, with Professor Johann Purkinje's crude classification process.

Almost a half a century later William Herschel began using fingerprints to identify pensioners in India, and Dr. Henry Faulds used finger prints in Japan to clear a burglary suspect (Herschel 1880).   Faulds attempted to start a fingerprinting bureau at Scotland Yard at his own expense but the offer was rejected in favor of Bertillonage (Saferstein 1990).[13]

Eventually Sir Francis Galton, building on Hershel's work, wrote the first book on fingerprinting, offering a systematic approach to the subject (Galton 1892). A few years later the police were using Edward Richard Henry's[14] classification system and fingerprints grew into a widely used identification system (Saferstein 1990).

Fingerprints at a crime scene are classified as Plastic, Patent or Latent. Plastic prints are those left in another substance (c.f. blood, wax, paint, etc). Almost any colored pigment can be used to fill in the fingermark impression.

---

[12] Edmond Locard (1877-1966) was the founder and director of the Institute of criminalistics at the University of Lyons in France.
[13]Bertillonage or anthropometry, was a system of body measurements developed by Alphonse Bertillon for the identification of criminals. It was eventually replaced by fingerprinting.
[14]Dr. Juan Vucetich had previously developed a crude fingerprint classification system but it was not effective for large numbers of prints.

Patent prints are those caused by transfer of some visible media (c.f., paint, ink, blood, etc.) on the fingertip to the surface impacted. This type of fingerprint is typically easy to locate and does not often need additional visualization aids. When they do need visualization, the method selected will be related to the nature of the transfer media.

Latent fingerprints are the often invisible impressions simply left by the finger. Fingertips contain glands which secrete a mixture of water, salt, oils and biological materials. These secretions coat the friction ridges and transfer the pattern to surfaces the finger touches. In order to process these latent prints, a variety of visualization and coloration methods have been developed.

One of common methods is the fuming of fingerprints with cyanoacrylate vapors.[15] As vaporized cyanoacrylate comes into contact with the secretions making the latent fingerprint it polymerizes into a colorless polycyanoacrylate, reproducing the pattern of the latent print in a more durable form. The next step then is to color the print so that it can be visualized and collected. Coloring of the polycyanoacrylate fingerprint, can be done in a variety of ways; the two most common being application of colored powders (pigments) or solution staining with dyes like Rhodamine 6G. Alternatively a colored polymer like CN-Yellow or Fuming Orange can be used in place of traditional cyanoacrylate.

The latent fingerprint can also be dusted directly with colored powders which will adhere to the residue left by the finger. A variety of colored pigments are used, but the most common is carbon black.

The background can also be colored with dyes not substantive to the fingerprint. Certain disperse dyes for example

---

[15] This technique has bee used in the US on non-porous evidence since the early 1980's when Ed German, a U.S. Army investigator, discovered his Japanese counterparts using the technique.

will color plastic and leave the fingerprint uncolored (Steele & Ball 2003).

The surface that bears the fingerprint is also important. For example, Axis Inversion Dyes which provide color alteration of fingerprints on polymers will obscure them if used on paper. Stains that work in combination with cyanoacrylate can destroy a naked print. It is important therefore to match the proper colorant with the proper application.

2.3 Colorant Selection

Laboratory methods and research papers will specify which colorants are to be used with the presented method. Often times these colorants were used because they were at hand from other methods or found through trial and error. Often times techniques will have only one associated colorant which as has been shown in the previous chapters limits their overall effectiveness.

The challenge then is to systematically be able to identify and select colorants which can expand the chromatic range of an applicable technique. A logical first step for this type of analysis is to examine the working colorants and then select alternate shades possessing similar properties. Fortunately here, most colorants have been used in science and industry for many years and a lot of information already exists to simplify the task of identifying the relevant properties.

Key Points

- Dyes and reactive chemicals can be used to locate and record blood pattern evidence.

- Chemiluminescence and fluorescent chemicals can be used to visualize faint blood traces in dark environments.

- Color changing chemicals can react with the background.

- Fingerprints can be Patent, Plastic of Latent.

- Colorants are useful for visualizing latent fingermarks.

- Colorant selection is dependent of the technique used.

References:

Barni F, Lewis SW, Berti A, Miskelly GM, Lago G. Forensic application of the luminol reaction as a presumptive test for latent blood detection. Talanta. 2007; 72(3):896-913.

Bodziak WJ. Use of leuco crystal violet to enhance shoe prints in blood. Forensic Sci. Int. 1996; 82:45-52.

Bremmer RH, de Bruin KG, van Gemert MJC, van Leeuwen TG, Aalders MCG. Forensic quest for age determination of bloodstains. Forensic Science International. 2011.

Cheeseman R, DiMeo LA. Fluorescein as a Field-Worthy Latent Bloodstain Detection System. Journal of Forensic Identification. 1995; 45(6):631-45.

Cheeseman R, Tomboc R. Fluorescein Technique Performance Study on Bloody Foot Trails. Journal of Forensic Identification. 2001; 51(1):16-27

Cheeseman R. Direct sensitivity comparison of fluorescein and luminol blood stain enhancement techniques. Journal of Forensic Identification. 1999; 49(3):8.

Cox M. A Study of the Sensitivity and Specificity of Four Presumptive Tests for Blood. *J. Foren. Sci.* 1991; **36:5**.

Galton, Francis; "Finger Prints" London: MacMillan and Co. (1892).

Grodsky M, W, K., Kirk, P.L. Simplified Preliminary Blood Testing. An Improved Technique and a Comparative Study of Methods. The Journal of Criminal Law, Criminology, and Police Science. 1951; 42(1):20.

Herschel, William James "Skin furrows of the hand". *Nature* 23 (578): 76. (1880).

Laux DL. Effects of Luminol on the Subsequent Analysis of Bloodstains. Journal of Forensic Sciences. 1991; 36(5):9.

Laufer, Berthold (1912). "History of the finger-print system". (1912).

Meyers TC, Cyril H, Rago JT. Forensic Science and Law: Investigative Applications in Criminal, Civil, and Family Justice. Boca Raton, FL, 2006: CRC Press.

Saferstein R. Criminalistics an introduction to forensic science. 4th ed. Englewood Cliffs: Prentice-Hall, Inc., 1990

Sahs PT. DAB (3,3'-Diaminobenzidine): An Advancement in Blood Print Detection. Journal of Forensic Identification. 1992; 42(5):9.

Steele, Charles, Ball, Mikki., Enhancing Contrast of Fingermarks on Plastic Tape, Journal of Forensic Science, Vol. 48, No. 6, November 2003.

Tontarski, K.L., Hoskins, K.A., Watkins, T. G., Brun-Conti, L., and Michaud, A.L., Chemical Enhancement Techniques of Bloodstain Patterns and DNA Recovery after Fire Exposure, J. Forensic Sci., 2009, 54:37-48

Chapter 3: Types of Colorants

While it is common to think of dyes used in blood detection as "blood detection dyes" and those used to visualize fingermarks as "fingerprinting dyes," these distinctions do not adequately describe the chemistries. Colorants can be categorized of as one of three types: Dyes, Pigments and Reactive Chemicals. Individual compounds can exist in more than one category simultaneously, and there are compounds which exhibit characteristics of a different category than the one they are in.

Forensic Scientists apply: Dyes, Pigments & Reactive chemicals to evidence to enhance visibility.

The first category are soluble direct colorants called "dyes." Dyes impart their color when they are dissolved into the material they are coloring. Typically, dyes provide a transparent color. So the color effect from dyes is additive with the ambient color of the material to which they are applied.

There are several different ways that dyes can be classifies. They can be grouped according to their: chemistry, shade, utility or industrial classification. For the purposes of this discussion, the most useful classification will be their industrial classification: Acid, Basic and Solvent (Including Disperse). Other dye classes exist but as a practical matter these three are the most important for analytical applications. (The Society of Dyers and Colourists. 1995)

Charge, chemistry, geometry, size and solubility are important factors determining whether or not a particular dye will provide color in a specific application. Industrial classifications take all of these factors into account but the real focus is on their utility. Dyes which tend to color similar substrates are grouped together.

Therefore, if one is looking for a starting point to expand the color options for a technique, they should begin with dyes of a similar industrial classification. Then within those categories, look for dyes with similar structure, charge, etc. To facilitate this approach, the dyes presented in this chapter are grouped by industrial classification.

3.1 Acid Dyes in Forensic Applications:

Acid dyes include colorants used in a wide range of applications including textile dyeing, lab staining and wood coloration. They are anionic, water soluble direct colorants. They are often salts of organic acids (Carboxylic, Phenol, or Sulfuric).

The name "Acid Dye" comes specifically from their use in textile applications. Acids are added to the dye bath to increase the number of protonated amino-groups increasing the affinity of fibers to form ionic bonds with the dye. Acid dyes are typically divided into three categories: Leveling, Milling and Metal complex.

Leveling dyes are typically of small molecular weight and prone to migration. While this limits their use in textile applications, it can make them useful for analytical applications like protein staining. The lower molecular weight also makes them well suited for color mixing allowing for targeted blends to maximize color differences.

Milling dyes are larger in molecular size. They are typically not as bright as leveling dyes and do not tend to be as vibrant in shade. They do not blend with other dyes well. But, they have greater resistance to being washed out once they have bonded with a cationic site.

Metalized dyes have been complexed with transition metals to form molecules with greater light and chemical resistance. These are commonly used in applications like wood stains where durability is of great importance.

Acid dyes in general can be chemically complex large aromatic molecules. Most acid dyes have sulfo or carboxy groups. The three main types are Anthraquinones, Azos and Triphenylmethanes. (AATCC 1997)

Anthraquinones are synthesized from a variety of intermediates. This is a common structure among blue leveling dyes.

Azo Dyes are based on azobenzenes. Azo dyes are often taken as an independent class due to regulatory concerns. Many red acid dyes have an azo structure. Azo dyes structures are also common in many different classes of water and solvent soluble dyes.

Triphenylmethanes are often the basis of acid milling dyes. They are frequently yellow or green shades.

While these chemistries are common to acid dyes, it is important to realize that these structures are also common to other dye classes as well. Although, as will be seen later, the color commonly generated by molecules using these structures can be different in the different dye classes.

On the following pages are some specific examples of common Acid dyes used in forensic applications.

Amido Black 10B

Image 14: Amido Black Data[14]

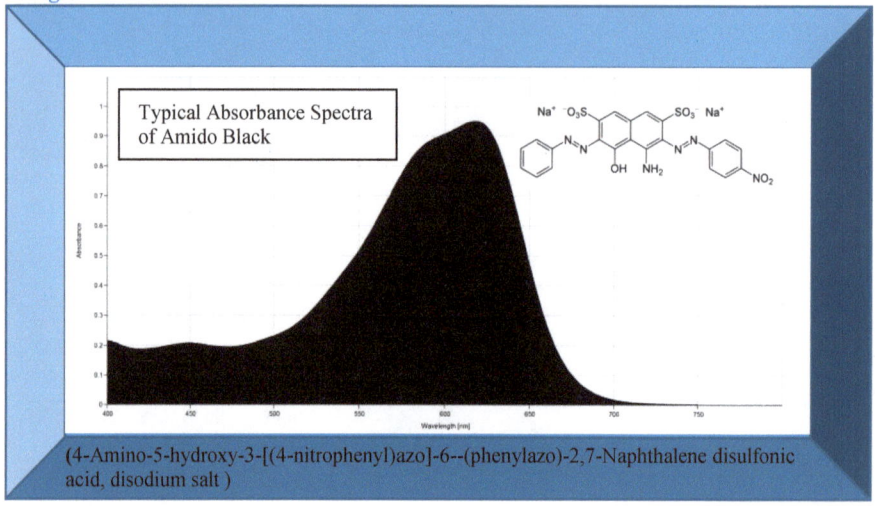

Typical Absorbance Spectra of Amido Black

(4-Amino-5-hydroxy-3-[(4-nitrophenyl)azo]-6--(phenylazo)-2,7-Naphthalene disulfonic acid, disodium salt)

CAS	1064-48-8
C.I.	20470

Amido Black (Acid Black 1) is an azo structure acid dye. Like other acid dyes, amido black is water soluble and has good affinity for bonding with cationic materials. Amido black also has alcohol solubility and is used in methanol preparations.

As can be seen in the absorbance curve above, although amido black has substantial absorption along the entire visible spectrum, but it absorbs most strongly between 540 nm and 640 nm. It produces a blue/black shade. In forensic applications it is used for blood staining and latent fingerprint detection.

When used to detect blood, Amido Black is prepared in an acidified methanol solution and applied directly to suspected dried blood. The excess is rinsed off and the dye will impart a dark blue color if blood or any suitable dyeable substrate is present. (cbdiai.org/Reagents/amidow)

---

[16] Optical measurement were collected using a Cary 3.

Blueblood™ Tracker

Image 15: Blueblood™ Tracker Data[15]

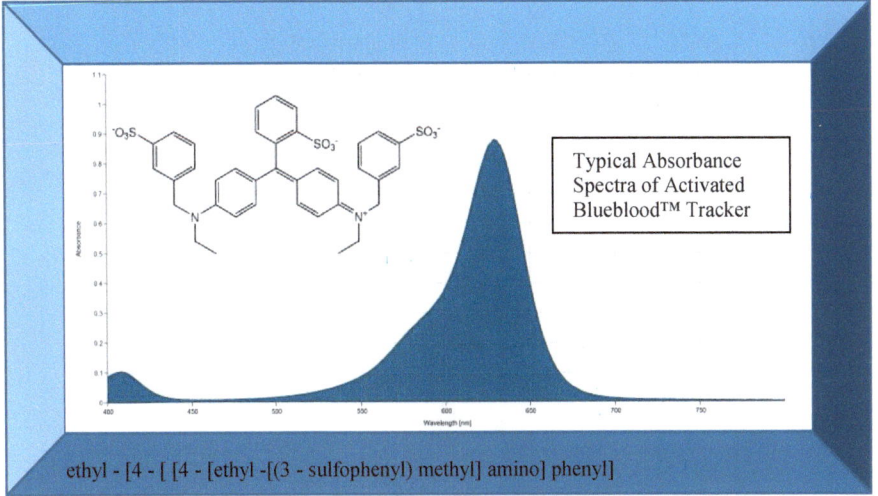

ethyl - [4 - [[4 - [ethyl -[(3 - sulfophenyl) methyl] amino] phenyl]

Typical Absorbance Spectra of Activated Blueblood™ Tracker

CAS	3844-45-9
C.I.	42090

Blueblood Tracker™ is the virtually colorless leuco form of Acid Blue 9 disodium salt. When it comes into contact with blood in the presence of an oxidizer it reacts to form the colored species.

Acid Blue 9 disodium salt is a common colorant used in a wide range of applications from cleaners to food coloring. In forensic applications it is used for blood stain detection. Once reacted, the acid blue 9 has a strong tinctorial value and because of the two significant absorption peaks near 410 nm and 630 nm, it is highly visible against a wide variety of backgrounds and allows for good light/dark discernibility for color blind users.

The color reaction is permanent so additional treatments are not required for subsequent viewing as is often the case with fluorescent and phosphorescent blood detection agents.

---

[17] Optical measurement were collected using a Cary 3.

Image 16: Top: Depletion set of bloody images
Bottom: Images treated with Blueblood ™ Enhanced

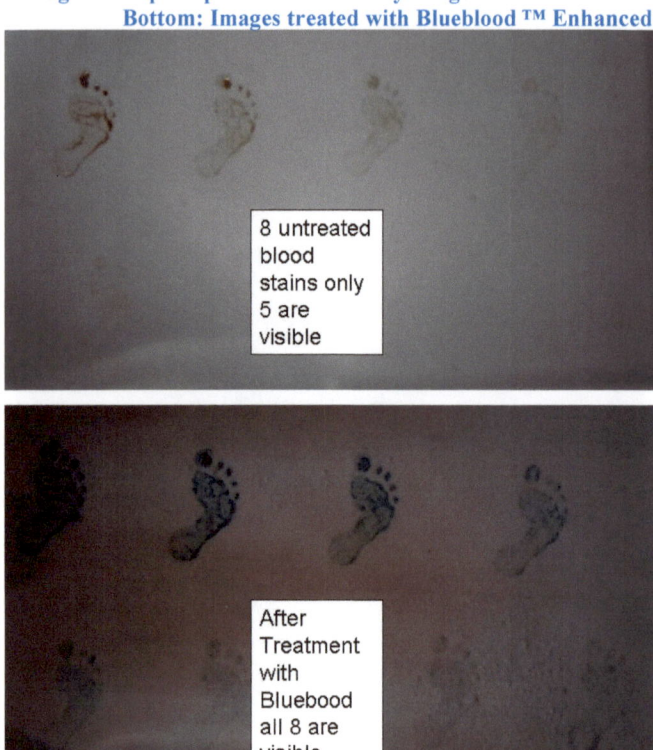

8 untreated blood stains only 5 are visible

After Treatment with Bluebood all 8 are visible

The Blueblood™ Tracker family of products currently includes three forms: Blueblood™ Enhanced, Blueblood™ Tracker and Blueblood™ Premix. All forms of the Blueblood™ Tracker are essentially colorless until they encounter the blood stain; then they turn blue.

For forensic applications, Blueblood™ Enhanced will both color and fix the blood stains. The Blueblood™ Tracker formula is for locating blood stains only. It is typically used by hunters and sportsmen. Both of these forms require mixing with an oxidizer prior to use. The premix form is a single component product which contains the oxidizer but has a more limited shelf life.

Similar to LCV and other blood detection colorants, Blueblood™ is an unreacted leuco dye that oxidizes in the presence of a developer and blood to form a colored species. Unlike LCV and many of the other leuco dyes on the market Blueblood is anionic rather than cationic and therefore has less biological affinity and is less reactive to other environmental contaminations. In addition, the anionic colorant that is formed in the reaction is acid blue 9, a common colorant used in a range of applications from cleaners to food coloring.

Crocein Scarlet 7b

Image 17: Crocein Scarlet Data[16]

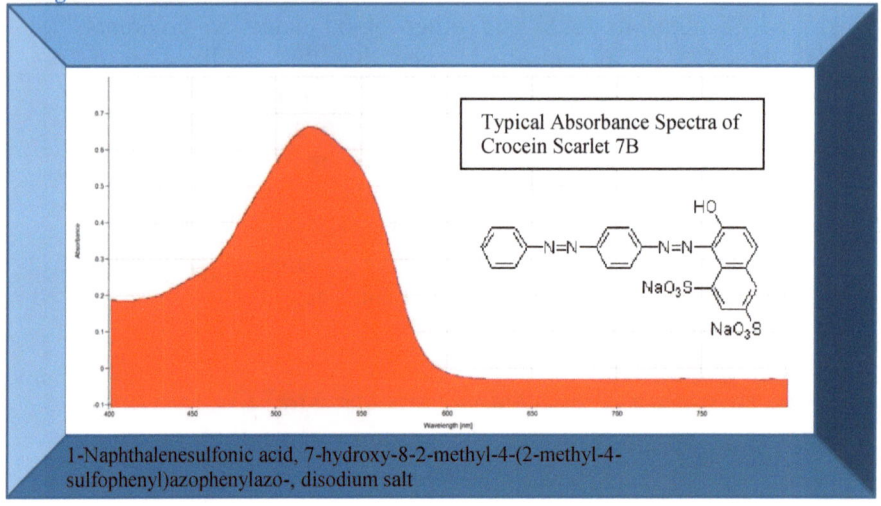

Typical Absorbance Spectra of Crocein Scarlet 7B

1-Naphthalenesulfonic acid, 7-hydroxy-8-2-methyl-4-(2-methyl-4-sulfophenyl)azophenylazo-, disodium salt

CAS	6226-76-2
C.I.	27165

Crocein Scarlet 7b (Acid Red 71) is used most commonly in dyeing applications. However its strong tinctorial value also makes it useful for biological staining. Depending on concentration the color will build from red to dark purple.

In forensic applications, the water soluble acid dye form is used in an acidified water solution as a stain. It imparts a weak color to the evidence but can be used in enhance contrast following other techniques. A solvent soluble derivative (Solvent Red 24) is also available.

---

[18] Optical measurement were collected using a Cary 3.

Fluorescein

Image 18: Fluorescein data[17]

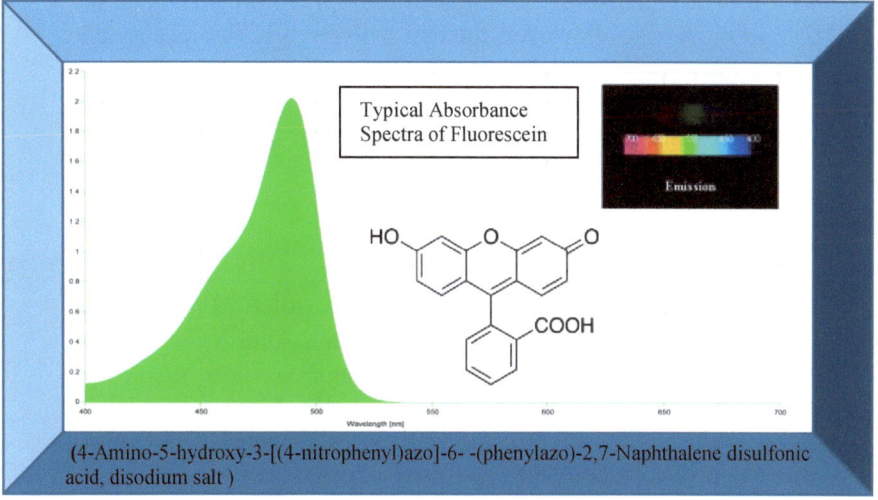

Typical Absorbance Spectra of Fluorescein

Emission

(4-Amino-5-hydroxy-3-[(4-nitrophenyl)azo]-6- -(phenylazo)-2,7-Naphthalene disulfonic acid, disodium salt)

CAS	2321-07-5
C.I.	45350

There are many forms of Fluorescein ranging from water to solvent solubility used in many applications from certified colors to industrial cleaners. What these varied forms have in common is that they are brightly fluorescent. As a result, Fluorescein is also used as a flow tracer and leak detector.

In forensic applications, fluorescein is typically used as blood detector. In its leuco form, it will react with blood in the presence of an oxidizer to form a fluorescent species.

Fluorescein and the associated dyes like Uranine have the distinct optical property of looking significantly different (changing from green to orange) when viewed via transmission, reflection or fluorescent emission.

In industrial applications, production inconsistencies will cause slight differences in the maximum excitation and emission

---

[19] Optical measurement were collected using a Cary 3.

of the fluorescence of fluorescein. It is therefore reasonable to expect that the relatively inconsistent reaction with blood residue will also produce some variation in the optical properties of the dye. In most cases however, the dye will strongly fluoresces near 520 nm when excited light between 385 nm and 500 nm.

Additional Fluorescent Acid Dyes for use with non-porous surfaces.

Acid Fuchsin (AKA Hungarian red, acid violet 19) is used to produce a violet stain in contact with blood. Acid Yellow 7 and acid yellow 17 can also be used on blood. These are water soluble, tinctorially weak colorants that will solubilize into blood stains. When used to stain blood they will fluoresce if excited with the proper light. (480 – 540 nm)

3.2 Basic Dyes in Forensic Applications:

Basic dyes are water-soluble, cationic colorants. They are typically very bright colorants with strong tinctorial effect. They have less chemical and light stability than acid dyes but are typically very good for biological staining applications. Basic dyes typically have amino or alkylamino groups. The use of basic dyes frequently requires the inclusion of an acid.

In addition to decorative applications like paper and flower dyeing, basic dyes are also use for textile dyeing of acrylic fibers. Basic dyes used in acrylic applications show great substantively and little or no migration (AATCC 1997).

Basic Yellow 40

Image 19: Basic Yellow 40 Data[18]

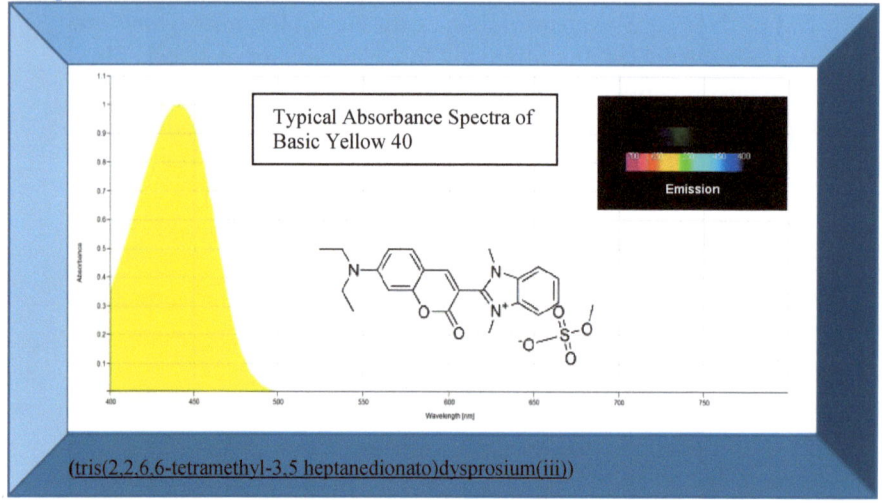

Typical Absorbance Spectra of Basic Yellow 40

Emission

(tris(2,2,6,6-tetramethyl-3,5 heptanedionato)dysprosium(iii))

CAS	29556-33-0
C.I.	12221- 86-2

Basic Yellow 40 is a water soluble coumarin fluorescent dye. It produces a yellow coloration and a brilliant greenish/yellow fluorescence under UV light. It is used in water and alcohol based applications like paper and inks. It is suitable for resin pigments, acrylic resins, and flexographic printing inks.

In forensic applications, it is used for blood staining and latent fingerprint detection. It is typically applied in a methanol solution, excited near 365 nm and viewed between 415 nm and 485 nm. (cbdiai.org/Reagents/by40)

However, as can be seen in the absorbance data above, Basic Yellow 40 absorbs most strongly between 420 nm and 480 nm and emits above 500 nm. Therefore strong visible light with yellow filters is also appropriate for viewing. Fluorescent

---

[20] Optical measurement were collected using a Cary 3.

photographic techniques for basic yellow 40 do recommend a 515 nm band pass filter.

Basic Yellow 40 is also sold under the name Maxilon® Flavine 10GFF.

Leuco Crystal Violet

Image 20: Leuco Crystal Violet Data[19]

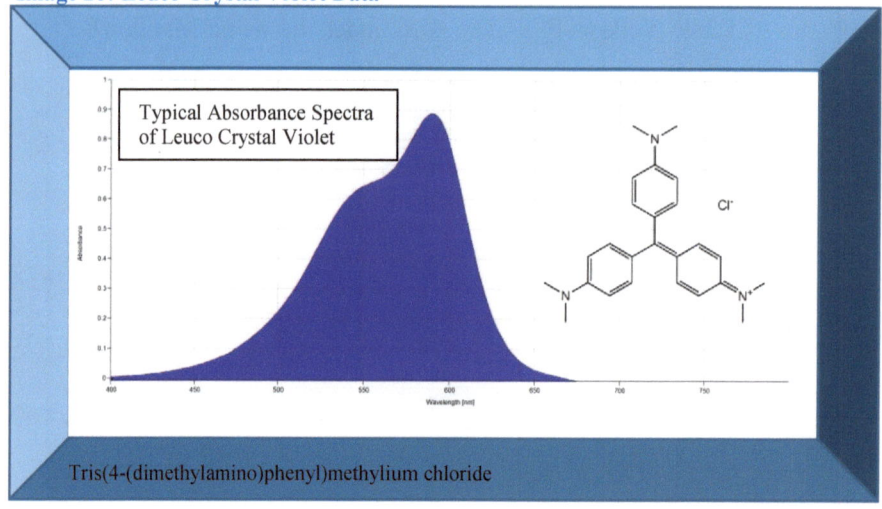

Typical Absorbance Spectra of Leuco Crystal Violet

Tris(4-(dimethylamino)phenyl)methylium chloride

CAS	584-62-9
C.I.	42555

Basic Violet 3 is available on the market under several different names (c.f. crystal violet, methyl violet, gentian violet etc.) It is a tinctorially strong water soluble cationic dye with excellent biologic affinity. Industrial applications include marking products where light fastness is not a requirement. In forensic applications, Basic Violet 3 is used in its colorless leuco form. When mixed with an oxidizer in the presence of blood (heme) it will react to form the colored species. (Fischer and Miller 1984)

---

[21] Optical measurement were collected using a Cary 3.

Malachite Green

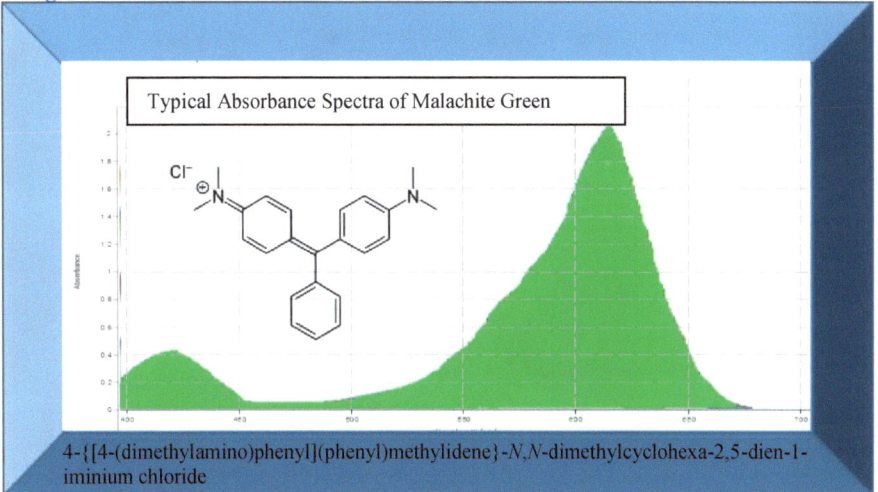

Typical Absorbance Spectra of Malachite Green

4-{[4-(dimethylamino)phenyl](phenyl)methylidene}-*N*,*N*-dimethylcyclohexa-2,5-dien-1-iminium chloride

CAS	584-62-9
C.I.	42555

Basic Green 4 (Malachite Green) is water soluble cationic dye used to dye paper, leather, grass and even pond water where it is used to control Ichthyophthirius multifiliis[23]. Several production methods for manufacturing the dye exist and the dye itself is available in solid form or in an acidic solution.

Regulatory concerns with this dye follow from a 1992 report where Canadian authorities listed it as a health risk. (Wendy 2006) This report however was related to a specific production method which is not commonly employed.

In forensic applications, Malachite Green is used as a counterstain, a laser dye and a blood detection dye. If the colorless leuco form is mixed with an oxidizer in the presence of blood (heme) it will react to form the colored species.

---

[22] Optical measurement were collected using a Cary 3.
[23] A disease common to fresh water fish.

Rhodamine 6G

Image 22: Rhodamine 6G Data[21]

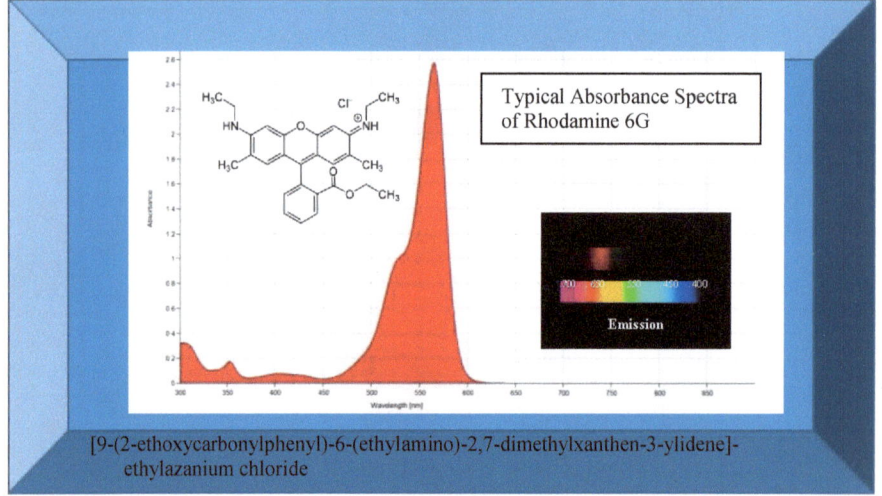

Typical Absorbance Spectra of Rhodamine 6G

Emission

[9-(2-ethoxycarbonylphenyl)-6-(ethylamino)-2,7-dimethylxanthen-3-ylidene]-ethylazanium chloride

CAS	989-38-8
C.I.	45160

Basic Red 1 (Rhodamine) is water soluble cationic dye used in ink and dyeing applications. It is one of many forms of rhodamine and replaces the more toxic Rhodamine B Base used in earlier biologic applications. Rhodamine 6G is marketed in three different chemistries.

---

[24] Optical measurement were collected using a Cary 3.

Rhodamine 6G chloride ($C_{27}H_{29}ClN_2O_3$) is a bronze powder. Although it is highly soluble, this version is corrosive to metals (except stainless steel.) The less soluble and less corrosive Rhodamine 6G Perchlorate ($C_{27}H_{29}ClN_2O_7$) is sold as red crystals. Rhodamine 6G tetrafluoroborate, ($C_{27}H_{29}BF_4N_2O_3$), appears as maroon crystals or powder. In any of its common forms, Rhodamine 6G is typically tinted with yellow (frequently basic yellow 40) to adjust the shade without sacrificing fluorescent strength. Solvent soluble forms of Rhodamine are also available.

Image 23:

Rhodamine stained CA prints chilled (L) and unchilled (R)

In forensic applications, Rhodamine 6G is used as a laser stain in fingerprinting. In this application the dye is dissolved in a methanol solution and washed over cyanoacrylate fumed prints. However, the dye also has a strong tinctorial value and can be used to visualize fingerprints directly. This is especially true when the evidence is cooled prior to fuming. (Steele 2012)

The image above shows two halves of the same sample. Both have been fumed with cyanoacrylate and washed with a rhodamine solution. The only difference processing is that the sample on the left was cooled first causing more irregularity in the formation of the polymer allowing for more rhodamine uptake.

Rhodamine has the advantage of being responsive to visible light, common lasers and inexpensive inspection lights.

Additional Basic Dyes used for used with cyanoacrylate fumed prints.

Basic Red 14 and Basic Red 28 have similar fluorescent emissions and can be used on cyanoacrylate fumed prints. These dyes are illuminated with a light source between 470 nm and 550 nm.

3.3 Solvent/Disperse Dyes in Forensic Applications:

Solvent dyes are typically non-polar or weakly polar. As such, they have little or no solubility in water. Typically they are used in hydrocarbon and non-polar solvents as well petroleum based products like fuels, waxes and polymers. Because of their lack of water solubility they are also used for pens and inks.

Like acid dyes, azo and anthraquinones are common structures. Among solvent dyes azo structures tend to be red or yellow. Anthraquinones tend to be green and blue.

Solvent dyes offer the same color options as their water based counterparts but the dyes have different substrate affinity. Solvent dyes tend to be better suited to coloring polymers and oils. The oil soluble dyes within this class tend to have good affinity to lipids.

When used in fingerprinting applications this allows them to be used to color latent fingermarks even when left by freshly washed skin. In addition, because they have little affinity to water, residue around the fingermark can be rinsed away without losing the color in the print itself.

The variety of available carrier solvents for solvent dyes also allows for different processing options and quicker drying times. High flash solvents can be used to deliver colorant quickly and be easily evaporated off. Low viscosity carriers can be used to drive dye into microscopic details and defects. Slow drying solvents can be used to increase the dwell time allowing dyes to migrate deeply into the evidence for increased visualization.

Some of the more common examples of solvent dyes used in forensic applications are presented on the pages following.

Nile Red

Image 24: Nile Red Data[22]

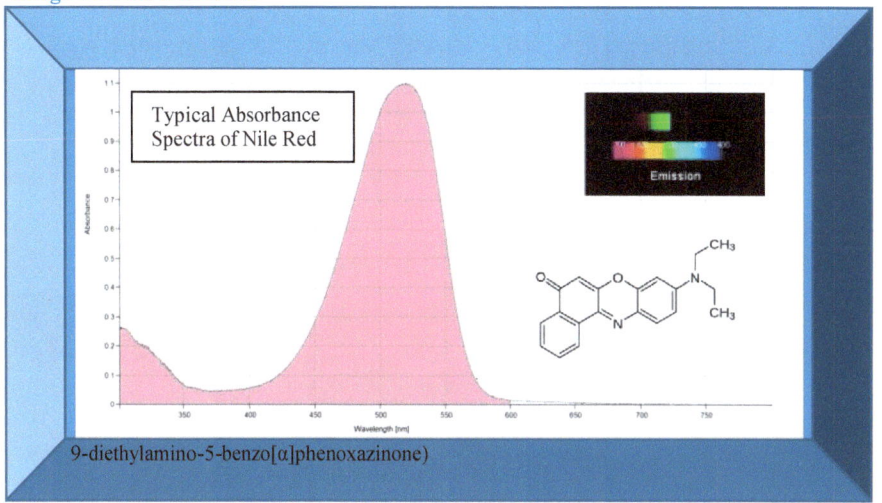

Typical Absorbance Spectra of Nile Red

Emission

9-diethylamino-5-benzo[α]phenoxazinone)

CAS	7385-67-3
C.I.	na

Nile Red (**Nile Blue Oxazone**) is a lipid stain derived from Nile Blue (Basic Blue 12). When Basic Blue 12 is boiled with sulfuric acid an amino group is exchanged for a carbonyl group. Its structure changes from an oxazine to an oxazone and it becomes solvent soluble. Nile red stains lipids red and can be easily visualized with an epifluorescence microscope. Nile Red will fluoresce under UV/Visible light from 360 - 480 nm and in lipid-rich environments with varying emissions from red to yellow depending on solvent. (See below)

---

[25] Optical measurement were collected using a Cary 3.

Nile Red is reported to excite from 360 nm to 485 nm, and emits over a wide range. This is heavily dependent on the solvent used. (Greenspan. Et.al. 1985)

Since the reaction to generate Nile Red does not usually completely exhaust the supply of Nile Blue, additional separation steps are required if pure Nile Red is needed.

Image 25: Solvent Based Color and Fluorescence of Nile Red

Top: Ambient Light   **Bottom:** Excited by 366nm
From Left to Right: Methanol, Ethanol, Acetonitrile, Dimethylformamide, Acetone, Ethylacetate, Dichloromethane, n-Hexane, Methyl-tert-Butyletherm Cyclohexane, Toluene

Oil Red O

Image 26: Oil Red O Data[23]

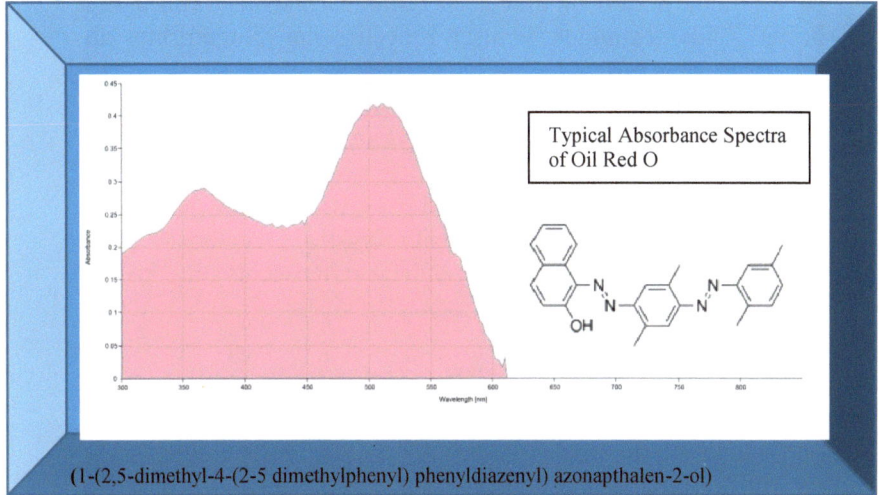

Typical Absorbance Spectra of Oil Red O

(1-(2,5-dimethyl-4-(2-5 dimethylphenyl) phenyldiazenyl) azonapthalen-2-ol)

CAS	1320-06-5
C.I.	26125

Oil Red O (Solvent Red 27) is an azo dye that produces a bluish-red color in a variety of plastic and petroleum solvent applications. It absorbs strongly until 600 nm and has a maximum absorption at 510 nm. The primary peak is between 467 nm and 550 nm.

Oil Red O is a lipid stain making fats more visible.[26] In forensic applications it is used to stain fingermarks on porous materials. (www.cbdiai.org, Triplett 2010, Beaudoin 2004). It mainly targets fat deposits on the surface of porous exhibits. It is a non-destructive technique and is a safe alternative to some of the Physical Developer methods. Oil Red O can also be used in sequence with other methods of fingerprints development. (Guigui & Beaudoin 2007)

---

[26] Optical measurement were collected using a Varian DMS 90.

In fingerprint applications, porous materials are immersed in a high pH methanol/water solution of Oil Red O for an extended period of time then removed and neutralized in a buffer. This technique is useful even when fingerprints are several months old.

Attention should be paid to the types of paper tested with this method. Oil Red is a solvent dye in an alcohol solution, as such certain wax fillers and gloss finishes may be susceptible to color uptake and provide undesired background color.

When used in sequence with Physical Developer testing, Oil Red O should be used first. (cbdiai.org/Reagents/oilredo)

Solvent Yellow 43

Image 27: Solvent Yellow 43 Data[24]

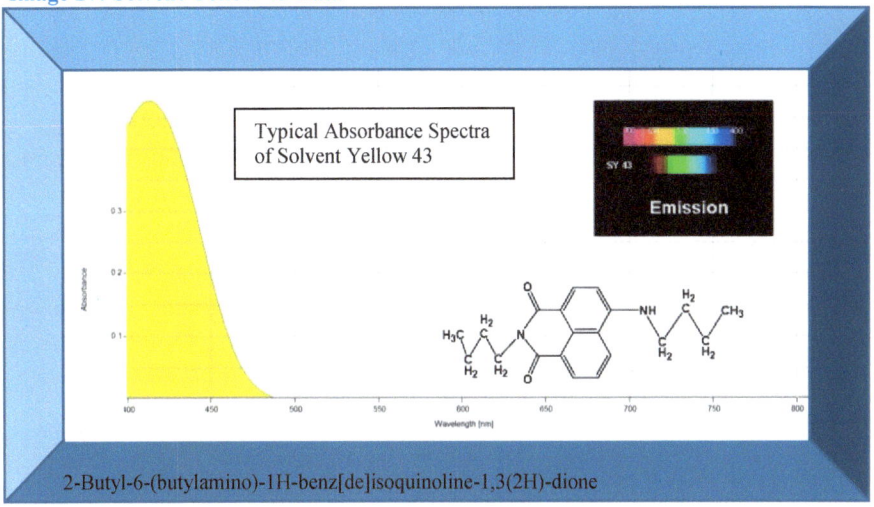

2-Butyl-6-(butylamino)-1H-benz[de]isoquinoline-1,3(2H)-dione

CAS	19125-99-6
C.I.	561930

Solvent yellow 43, also known as Naphthalimide yellow, is a solvent soluble fluorescent dye that has industrial uses ranging from smoke colorants to plastic dyeing. (Steele 2006) While it is tincitorially weak, it has a strong fluorescent emission from 530 nm to 600 nm when excited between 438 nm and 480 nm.

Solvent yellow 43 is also used as a tracer dye in non-destructive testing products similar to Ardrox. (Steele 2000)

In forensic applications solvent yellow 43 is used in penetrating fluids to invade the cured polycyanoacrylate post fuming. The dye is then excitable with a UV light.

---

[27] Optical measurement were collected using a Cary 3.

Solvent Yellow 43 has also been liganded to cyanoacrylate a monomer to form products like CN-Yellow. This colored monomer can be used in place of cyanoacrylate to fume fingerprints and directly provide both the visual color of solvent yellow 43 and the fluorescent emission.

Image 28:

Fingermarks visualized with CN -Yellow

CN-Yellow offers an alternative to traditional cyanoacrylate fuming that provides resin and color in one step eliminating the need for subsequent dye staining. In addition, because CN-Yellow fluoresces, fingerprint development can be observed in real time diminishing the chance for overdevelopment of the print.

CN-Yellow is offered in fuming wand cartridges and solid crystals. Typical usage is 2.5g of CN-Yellow Crystals per 10 cubic foot volume to be fumed.

It should be noted that the crystals sublime at approximately 480°F so high temperature hot plates are needed.

Fuming Orange

Image 29: Fuming Orange Data[25]

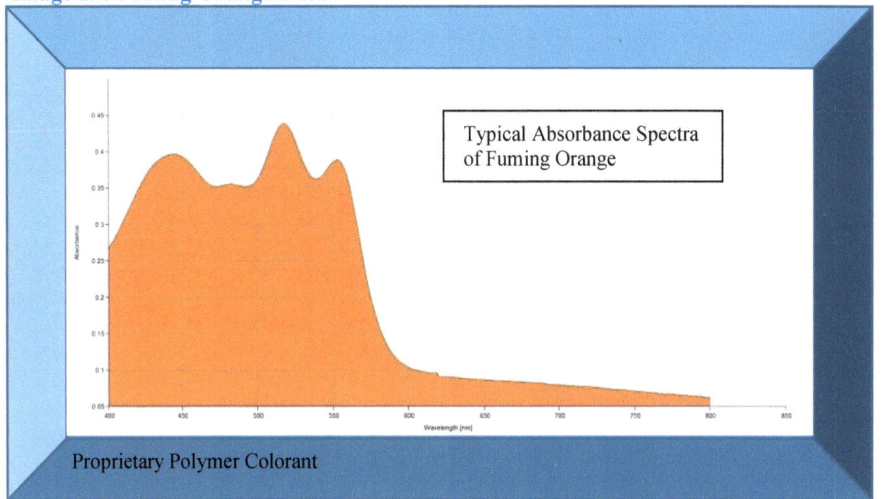

Typical Absorbance Spectra
of Fuming Orange

Proprietary Polymer Colorant

Although it is a finished resin not simply a colorant, Fuming orange is mentioned here because it is similar to CN-Yellow. It can also be used in place of traditional cyanoacrylate fuming to provide a fluorescent/colored print in a single step.

Image 30: Fuming Orange Fingerprints

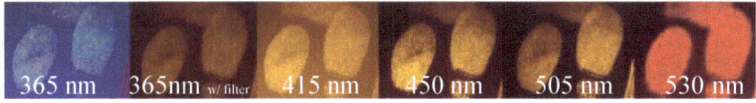

365 nm   365nm w/ filter   415 nm   450 nm   505 nm   530 nm

Fingerprint processed with Fuming Orange and viewed under different wavelengths of incident light

Show in the image above, the Fuming Orange resin fluoresces under excitation from a wide variety of lights sources, and like CN-Yellow reduces the risk of overdevelopment.

---

[28] Optical measurement were collected using a Cary 3.

In addition, a comparative analysis determined that 89% of latent finger prints developed with Fuming Orange were deemed identifiable as opposed to 70% processed with the Rhodamine 6G method. Shown at the side is a latent fingermark split in half.

Image 31: Fuming Orange vs. Rhodamine

Finger print split in two.
L: Processed with Fuming Orange
R: Processed with Rhodamine 6G

The left side of the latent fingermark was processed with Fuming Orange and the right side was processed with Rhodamine 6G after cyanoacrylate fuming and according to standard methodologies. As can be seen, the print processed with the Fuming Orange produced more even coloration.

When a dye stain is applied to a traditionally processed cyanoacrylate print, color deposition is weaker or stronger in different parts based on surface shape and resin structure. By contrast because the Fuming Orange resin itself is fluorescent the color is more uniform making detail easier to read. On a scale of 0 (no detail) to 4 (best detail) Fuming Orange samples had a mean score of 3.48 and the mean for Rhodamine 6G samples was 2.94. (Hines 2012)

3.3.1 Axis Inversion Dyes:

Disperse dyes are a special class of solvent dyes. They are water insoluble colorants with an affinity for various polymer substrates. (AATCC 1997). Although they are insoluble in water they are frequently blended with dispersing aids to allow them to be used in water applications.

Within this general class of dyes is the subset of sublimation or heat transfer dyes. These dyes provide strong visual color by a sublimation process, which functions differently than the traditional fuming methods familiar to fingerprint examiners. Sublimation Dyes typically have small molecular weight and with heat or proper solvents can be made to invade a variety of substrates.

When sublimation dyes are properly heated, the resulting vapor invades materials like polymers and once in the material it crystallizes to form a colored component. This method is used to print polyester, produce colored smoke and perform fume based dyeing and staining (AATCC 1997).

In forensic applications, a laboratory grade of sublimation dyes known as Axis Inversion Dyes (AXI), can be used to color the background around a fingermark left on a polymer substrate like plastic tape. (Steel & Ball 2003)

Penetration of the AXI dyes into polymer surfaces is blocked by latent fingerprint residue. As a result the AXI dyes impart color into the polymer everywhere except were the latent fingerprint is blocking it. The effect produces a "negative image" of the fingerprint. While this can be confusing for the examiner who is used to working with positive images, AXI dyes have the advantage that the image will remain even after the fingerprint itself has faded away.

There are three colors of AXI dyes available; Blue, Red and Yellow. In addition the AXI dyes are all of similar size and weight and blend easily. They can therefore be blended to form a full range of colors to be used against a variety of backgrounds.

Axis Inversion Blue

Image 32: Axis Inversion Blue Data[26]

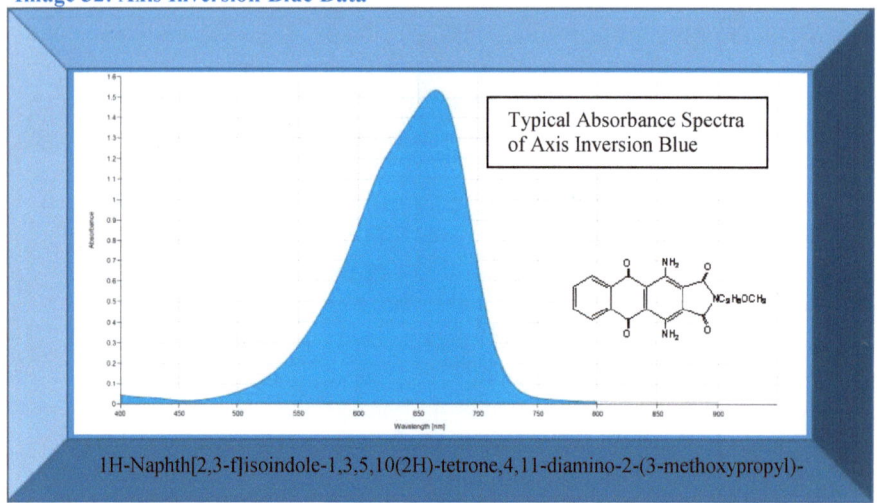

Typical Absorbance Spectra of Axis Inversion Blue

1H-Naphth[2,3-f]isoindole-1,3,5,10(2H)-tetrone,4,11-diamino-2-(3-methoxypropyl)-

CAS	12217-80-0
C.I.	61104

Axis Inversion Blue (AXI Blue) is a laboratory grade of disperse blue 60. It sublimes at temperatures above 200°C. When it crystalizes to form a colored particle, it absorbs light between 450 - 650 nm and reflects photons betweem 400 - 450 nm and 650 - 700 nm. AXI blue is therefore appropriate for developing latent fingermarks on materials with primary color along the green-red axis like brown mailing tape (3M cat 143). (Steele & Ball 2003)

When developing fingermarks on brown tape, the forensic scientist can either fume the tape with the dye directly or in combination with cyanoacrylate. When uses by itself AXI Blue has been shown to have good affinity for both the smooth and adhesive sides of the tape, except where the fingerprint was deposited. In both cases, the fingerprint acted like a resist and

---

[29] Optical measurement were collected using a Cary 3.

prevented the dye from penetrating the tape underneath. The effect was to visualize the fingerprints by coloring the background.

When fuming with AXI blue in combination with cyanoacrylate, the dye can be sublimed at the same time or in sequence after the sample has been fumed with cyanoacrylate. The best results have been shown to be obtained with the sequential approach.

Axis Inversion Red

Image 33: Axis Inversion Red Data[27]

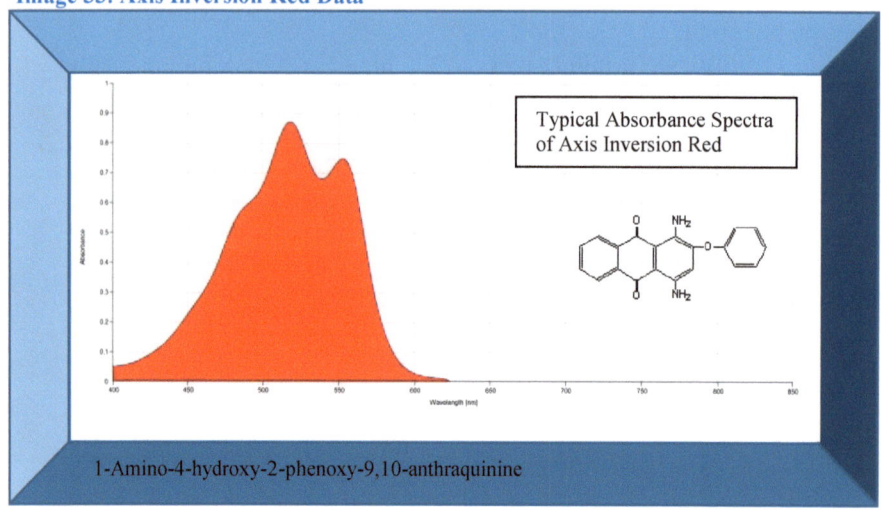

Typical Absorbance Spectra of Axis Inversion Red

1-Amino-4-hydroxy-2-phenoxy-9,10-anthraquinine

CAS	17418-58-5
C.I.	60756

Axis Inversion Red (AXI Red) is a laboratory grade of Disperse Red 60. In this form Disperse Red 60 is a sublimation dye that can be used for a variety of applications ranging from smoke dyes to textile printing.

In forensic applications, AXI red is appropriate for processing evidence with primary color development along the yellow-blue axis like Low Density Polyethylene (LDPE) bags, water bottles and other clear plastic evidence types.

The coloristic characteristics of clear resins can be confusing when deciding relevant color axis. Although the objects may appear clear, in truth they are not. Clear polymers tend to yellow in processing and over the life of the manufactured products. This yellowing is countered with a

---

[30] Optical measurement were collected using a Cary 3.

bluing agent of some sort to give the appearance of clarity. AXI Red provides coloration along the perpendicular axis

Because it is a subliming colorant, AXI Red has some advantages and disadvantages over traditional latent fingermark processing methods. Among the advantages are: ease of processing, increased visibility vs. cyanoacrylate and storage stability. The disadvantages are the need for fuming equipment and decreased visibility vs. Carbon Black.

For samples like LDPE bags, which are commonly encountered in drug arrests and which can leave the laboratory with a large number of samples to process AXI Red provides a coloristic alternative to traditional methods. In these cases, processing time, visibility and storage stability can be critical issues.

The traditional choices of Carbon Black Dusting or Cyanoacrylate fuming make the forensic scientist choose between visibility and speed. AXI red offers an intermediate option. Methods using AXI red have been shown to have the speed of CA fuming making it significantly faster than dusting with carbon black with only a slight sacrifice of visibility.

As stated earlier, the human eye can typically detect color shifts of DE >1 on the CIELab scale. Carbon black on LDPE bags produces the strongest shift with an average DE of greater than 20. This number is so high as to be beyond a reasonable range of comparison.

Images produced with AXI Red are also well above the DE >1 threshold with a DE of approximately 10. While this is measurably less than the color difference produced with carbon black, it is still high enough to be beyond a reasonable range of comparison.

Perhaps the more significant comparison comes from a blind study where viewers rated fingerprints generated with AXI Red on a visibility scale of 1-4 as an average 2.17, only slightly below the 2.38 that carbon black was rated.

In addition to good visibility, latent fingermarks developed with AXI red are more stable than CA fumed prints or Carbon Black dusted prints.

Image 34: AXI Red on LDPE

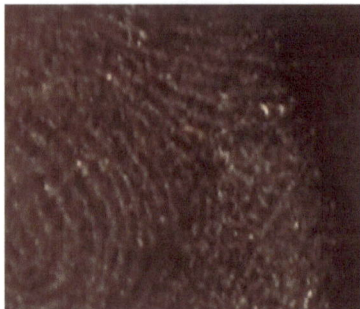

Fingerprint on LDPE processed with AXI red after 30 days in storage

Tradition methods tend to focus on coloring the fingermark. AXI dyes color the background. As a result images generated with AXI colors have great stability. In a study at the University of Illinois at Chicago, none of the fingermarks developed with the AXI red were damaged or altered by the various environmental conditions. Even high heat and high humidity environment had no impact on fingermarks developed with AXI red.

The image shown above is of a sample stored for thirty days at 40°C[31] and 80% humidity. After this length of time no deterioration was found. By comparison, during the same period, deterioration was seen in sample processed with cyanoacrylate and carbon black. (Steele 2014)

---

[31] The high heat and high humidity samples were run at 40°C. Based on the performance range of the incubator used the practical range of variation was of 35°C to 45°C. It should be noted that above 50°C. LDPE bags begin to physically distort.

Axis Inversion Yellow

Image 35: Axis Inversion Yellow Data[29]

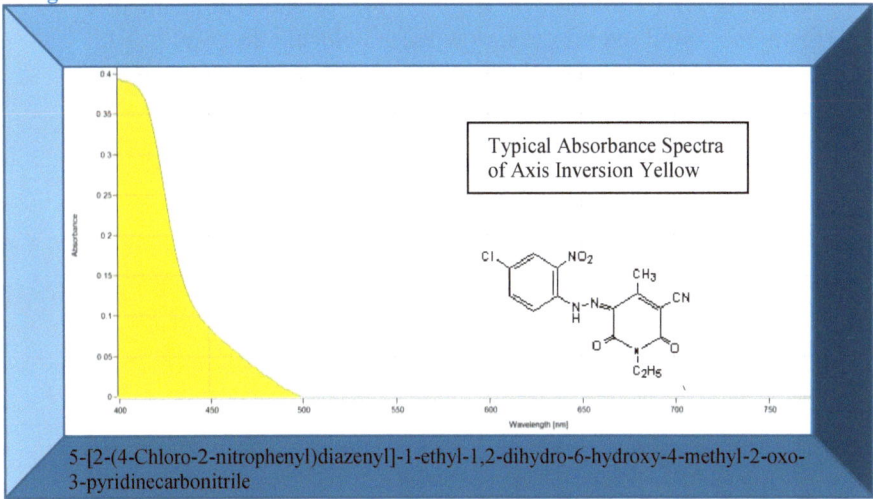

Typical Absorbance Spectra of Axis Inversion Yellow

5-[2-(4-Chloro-2-nitrophenyl)diazenyl]-1-ethyl-1,2-dihydro-6-hydroxy-4-methyl-2-oxo-3-pyridinecarbonitrile

CAS	86836-02-4
C.I.	12755

Axis Inversion Yellow (AXI Yellow) is a laboratory grade of Disperse Yellow 211 that provides coloration against black surfaces like electrical tape and gel lifters. (Steele& Ball 2003, Steele 2013)

Fingermarks, shoe impressions and other pattern evidence are often cluttered with background dirt or concealed among layers of other impressions. When these conditions are encountered, gel lifters can be used to gently remove one layer of residue at a time from a surface. The impression created by the residue on the gel surface is faint, but a variety of techniques, lighting options and imagers can resolve clear usable impressions.

For those that do not have access to specific imagers for gel lifters, the impression can also be enhanced with the use of

---

[32] Optical measurement were collected using a Cary 3.

colorants. As with all contrast aids the proper color is essential to accent the image. For black gel lifters, a variety of yellow dyes, including AXI Yellow will add 1-3 units of shade adjustment to the area around the impression without coloring the print itself.

The process works like any other sublimation printing process. The dye is heated to its sublimation temperature. Gel lifters are very susceptible to heat deformity. So, heat the dye away from the gel lifter and direct the colored vapor across the surface.

Image 36: Gel Lifter

Fingerprint on BVDA Gel Lifter
post fuming with AXI Yellow

When the vapor impacts the gel lifter surface some dye solubilizes into the material. The lifted impression however, works like a printer's resist preventing the dye from coloring the gel lifter under the impression surface.

The result is that the impression is differentiated from the background without the need for expensive equipment.

This technique is compatible with most existing imaging techniques and only serves to improve the image results. (Steele 2013)

3.4 Reactive Chemicals in Forensic Applications:

In addition to the various lueco dyes used to produce color, many other dye intermediates and reactive chemicals will also produce color when oxidized in the presence of lipids and heme. Some common examples are presented herein.

ABTS

2,2'-azino-di-[3-ehtylbenzthiazolinesulfonate(6)]diammonium salt

CAS	30931-67-0
C.I.	na

Image 37

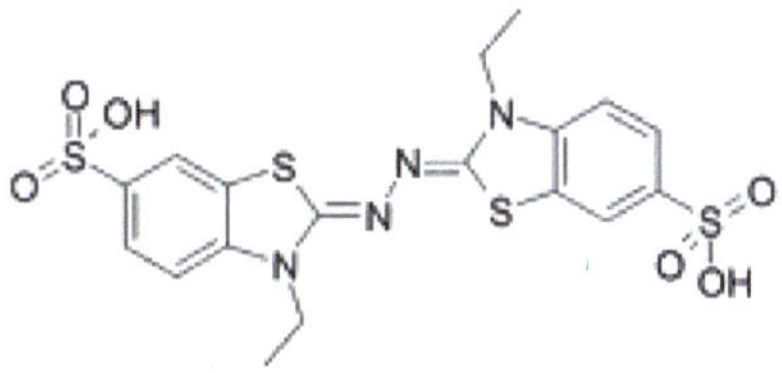

ABTS is an enzyme reactive chemical that develops a green color in certain reactions. It used commonly in food and agricultural industries to allow for spectrophotometric monitoring of reactions.

In forensic applications, it reacts in the presence of an oxidizer when in contact with blood and to from a green colorant. It is used on both porous and non-porous surfaces. This reaction can be used after treatment with Ninhydrin.

Silver Nitrate

CAS	7761-88-8
C.I.	na

Image 38

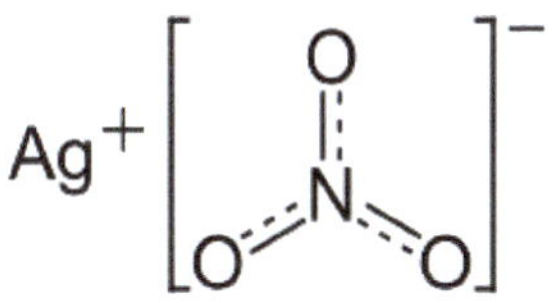

Silver Nitrate is used on its own and as a part of the Physical Developer process. Silver Nitrate is predominantly used as a biological stain.

It will react to form a brown colored species in the presence of a variety of biological material. In forensic applications, Silver Nitrate is used to evolve fingerprints on porous materials. Weak solutions of 1%-3% water or alcohol can be applied to paper and the like to produce a brown colored fingerprints.

Silver Nitrate bonds to the salts left by the latent fingermarks on porous materials when reacted with UV (365 nm light) (Natural or Artificial). Silver Nitrate can react with many common background components. It therefore does produce a significant background staining.

Silver Nitrate should be used after staining with other reagents like Ninhydrin. (cbdiai/Reagents/agno) It should also be noted that the brown color may not be the best option for dark, brown or red backgrounds.

Similarly the Physical Developer process uses chemical solutions and a silver nitrate dispersion to form a brown to black finger print on porous surfaces.

DIAMINOBENZIDINE
3,3'-Diaminobenzidine tetrahydrochloride

Image 39

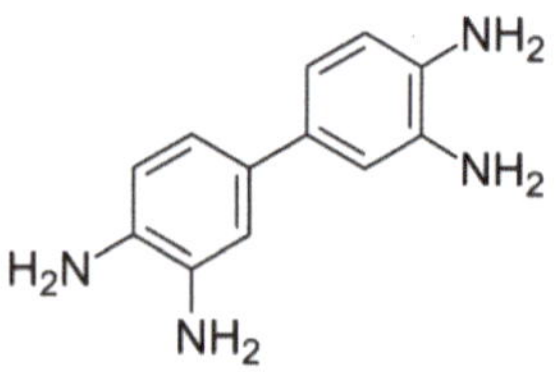

DIAMINOBENZIDINE (DAB) reacts to form a brown color in the presence of an oxidizer and the heme groups in blood. It is most commonly used to detect bloody fingerprints on porous and non-porous surfaces alone and in combination with Amido Black and ABTS. DAB isn't affected by most detection chemistries, but will of course be block if used after resonating processes like fuming with cyanoacrylate or Fuming Orange. DAB must therefore be used before resin fuming. (cbdiai.org/Reagents/dab)

Ninhydrin:
2,2-dihydroxy-1,3-indandione

Image 40

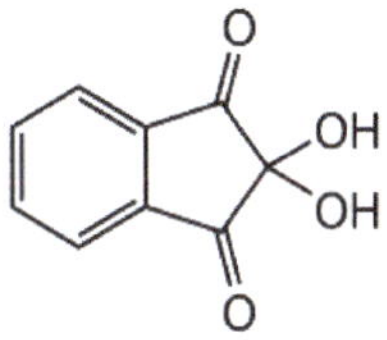

Ninhydrin is a solvent soluble chemical that reacts with amino acids to form Ruhemann's Purple.[33] It has been used to detect fingermarks on porous surfaces to detect latent fingerprints since 1954. (Oden & von Hofsten, 1954)

Ninhydrin is typically prepared in a high flash solvent so that it dries quickly.

Image 41

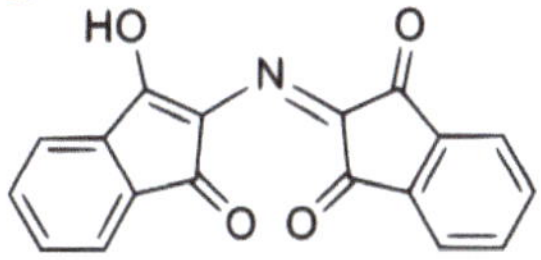

Treatment with ninhydrin can be done after the evidence has been treated with DFO or iodine. If ninhydrin is used in combination with silver nitrate or physical developer treatment it must be done first.

Ninhydrin solutions are typically sprayed onto the evidence which is then dried. Once dried it can be placed under heat in a humid environment 60%-70% to facilitate the reaction forming the Ruhemann's purple.

DFO
1,8-Diazafluoren-9-one

Image 42

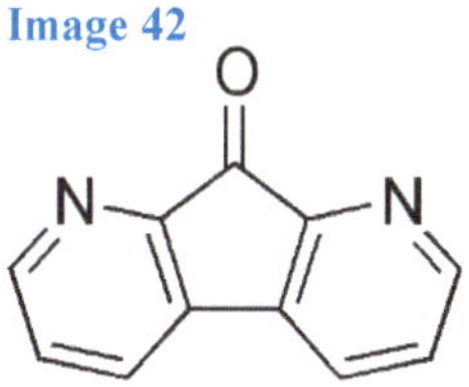

DFO was introduced to the forensic community in 1990. Like ninhydrin it reacts to form its colored species in the presences of Amino acids. When reacted it produces a weak visual color and fluorescent emission. The fluorescence is excited near 470 nm and emits near 570 nm. It can be used on porous surfaces prior to ninhydrin applications to yield better results. (Pounds & Grigg 1990)

Iodine

---

[33] So named because it was first discovered by Siegfried Ruhemann, in 1910.

Iodine fuming is essentially a non-destructive test method. Elemental Iodine will sublime at room temperature[34] and temporarily invade fats and a variety of materials allowing the photographing of images like fingerprints. Iodine be fumed across a variety of surfaces including porous surfaces to produce yellow/orange to pink purple images.

Image 43

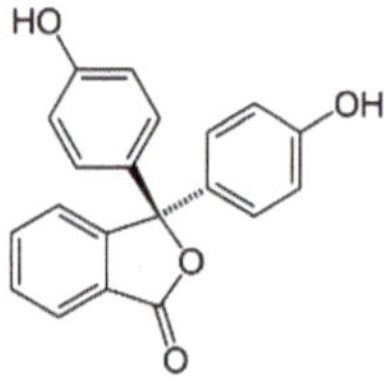

Phenolphthalein
3,3-bis(4-hydroxyphenyl)isobenzofuran-1(3*H*)-one

Phenolphthalein is a pH indicator used to detect the presence of blood. It is colorless in acidic conditions and moves through various shades of pink as the pH rises.

---

[34] Iodine has a vapor pressure of 0.3 mm at 25°C.

3.5 Pigments in Forensic Applications:

Pigments are insoluble direct colorants which, impart their color when they are applied to a material. Typically, pigments provide an opaque color. So the color will cover the existing color of the material to which it is applied.

As with dyes there are several different ways to categorize pigments. For the purposes of most forensic applications stability to processing environments is not an issue. Pigments can be thought of therefore as simply organic or inorganic and submicron or standard.

In forensic applications, pigments are typically uses to fill impressions or adhere to residue. Pigments only interact with the sample physically, that is, there is no chemical effect. As such, only the size, geometry and color are of significant value.

Pigments can be purchased in virtually any color, they can be magnetic, and some have the inclusion of fluorescent components to further enhance visibility. But the most common pigment used is carbon black. Because this is a true black it is visible against almost any background.

Carbon Black

Image 44: Carbon Black Data[32]

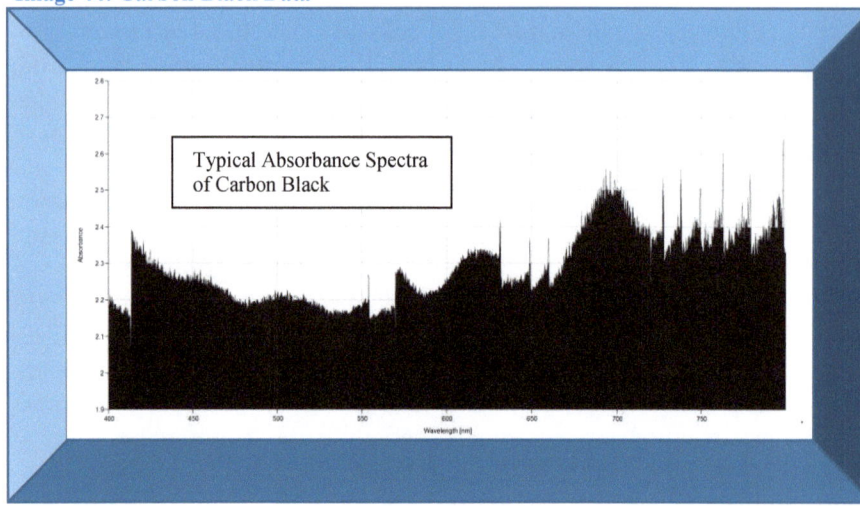

Typical Absorbance Spectra of Carbon Black

CAS	1333-86-4
C.I.	77266

Carbon Black is produced by burning hydrocarbons like oil or natural gas. The fuel type is dictated by the specific process and also defines the classification of resulting particle. Furnace Black, is produced by burning petroleum or coal oil. Channel Black is produced by burning natural gas. Lamp black is produced by burning oils and woods. Etc.

Channel Black has the highest level of purity making it ideal for forensic applications if ground to a small enough size. Furnace Black is easy to manufacture and allows for some control of particle size and also has forensic applications.

The "fundamental" particle size of furnace black is 10 to 80 nm, however, the reality is that van der Waals forces bind these particles into solid agglomerates up to 5,000 nm or greater

---

[35] Optical measurement were collected using a Cary 3.

in size. (Surinder 2010, Steele et.al. 2011) These particles can be broken up by grinding or separated out in production.

In forensic applications, small particle carbon black is dusted on fingerprints either as they are or post cyanoacrylate fuming to allow for a dark colored image of the print which can be lifted from the existing surface or photographed.

In addition, a more durable option of coloration using carbon black also exists. When produced by flame, the newly formed particles are carried by thermal convection currents. Since heavier/larger particles will not travel as far smaller/lighter ones, particle size can be controlled by proper placement of the collection vehicle. This allows for the formation of specifically sized particles. When used in combination with cyanoacrylate fuming, dark black fingerprints can be developed. (Steele et.al 2011)

Key Points

- There are many types of colorants.

- Colorants can be grouped by their industrial class.

- Color spectrum data is available for most colorants.

References:

American Association of Textile Chemists and Colorists. Technical manual of the American Association of Textile Chemists and Colorists. Vol. 72. The Association, 1997

Beaudoin, A. New technique for revealing latent fingerprints on wet, porous surfaces: Oil Red O. Journal of Forensic Identification, 2004, 54 (4), 413-421.

Fischer, J. F. and Miller, W. G. (1984) 'The Enhancement of Blood Prints by Chemical Methods and Laser Induced Fluorescence', Ident. News, vol. 34 (7), p

Greenspan, E. P. Mayer and S. D. Fowler "Nile Red, A Selective Fluorescent Stain for Intracellular Lipid Droplets". *Journal of Cell Biology* **100** (1): 965–973.) (1985).

Guigui, K.; Beaudoin, A. The use of Oil Red O in sequence with other methods of fingerprint development. Journal of Forensic Identification, 2007, 57 (4), 550-581.

Hines, Mason: *A Statistical Comparison of "Fuming Orange" Latent Fingerprint Developer to Traditional Dye Staining with Rhodamine 6G.* Evidence Technology Magazine, September-October 2012

http://www.cbdiai.org/Reagents/amidow.html

http://www.cbdiai.org/Reagents/by40.html

http://www.cbdiai/Reagents/agno

http://www.cbdiai.org/Reagents/oilredo.html

http://www.cbdiai.org/Reagents/dab.html

Oden, S. and B. von Hofsten, Detection of fingerprints by the ninhydrin reaction. Nature, 173 (1954):449.

Pounds, C.A., R. Grigg, T. Mongkolaussavaratana, The use of 1,8-diazafluoren-9-one (DFO) for the fluorescent detection of latent fingerprints on paper: a preliminary evaluation. Journal of Forensic Sciences, 35 (1990):169–175.

Steele, Charles A. Improving the Stability of Stored Fingermarks on Plastic Bags by Axis Inversion Development http://hdl.handle.net/10027/18771 May 2014

Steele, Charles: *Forensic Concepts: Enhancing Lifters, Aneval Inc. 2013*

Steele, Charles; et.al. Forced Condensation of Cyanoacrylate with Temperature Control of the Evidence Surface to Modify Polymer Formation and Improve Fingerprint Detection/Visualization, Journal of Forensic Identification, July/August 2012, Vol. 62, No. 4

Steele, Charles et.al. Specific Heat Capacity Thermal Function of the Cyanoacrylate Fingerprint Development Process NIJ 2009-DN-BX-K196 2011

Steele Charles, (ed.) Technical Bulletin 57: Keystone Smoke Dyes, Keystone Aniline Corporation March 29, 20006

Steele Charles, (ed.) Technical Bulleting 100: Flaw and Strain Detection, Keystone Aniline Corporation August 18, 2000

Steele, Charles, Ball, Mikki., Enhancing Contrast of Fingermarks on Plastic Tape, Journal of Forensic Science, Vol. 48, No. 6, November 2003.

Surinder, P. Petroleum Fuels Manufacturing Hand Book, Chapter 9.McGraw-Hill, United States 2010

The Society of Dyers and Colourists. American Association of Textile Chemists and Colorists colour index international. 3rd Ed. The Society, 1995

Triplett M, *Fingerprint Dictionary*, Two Rings Publishing, Bellevue, Washington. 2010

Wendy C. Andersen, Sherri B. Turnipseed, and José E. Roybal "Quantitative and Confirmatory Analyses of Malachite Green and Leuco Malachite Green Residues in Fish and Shrimp" *J. Agric. Food Chem.* 2006, volume 54, pp. 4517–4523.

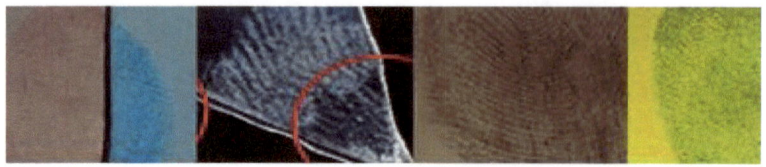

Chapter 4: Applications of Color Science

A sample may be the color of the background, so faint it cannot be seen or simply uncolored. In these cases a variety of techniques have been developed to color the evidence. These techniques use many different chemistries to achieve the desired visualization. Often, more than one option for providing color exists. In fact, as can be seen in the information above, often multiple products are available which impart virtually the same color under the same conditions.

The examiner should therefore take into account both the ambient color, and also the specific absorption or reflection of the desired target. Armed with that information, the specific colorant which best alters the desired axis can be selected.

A basic rules of thumb that can be applied when selecting colorants is to adjust an axis perpendicular to the primary color axis of the ambient shade. That is to say if the primary color axis is yellow-blue then adjust the shade along the red-green axis. The trick is to know what the ambient dominant axis is.

Image 45

Absorbance Spectra of Mahogany Wood Stain

The spectrum above shows a typical mahogany wood stain. It has absorption peaks near 460 and 630 nm. This means that it is predominantly absorbing blue and orange photons. So, the color impact of the blue/yellow axis is less significant to the eye of the observer than the red/green axis. Adjusting the red/green axis therefore will have less impact on visualization of evidence than adjusting the blue/yellow axis.

For example, if an examiner was looking for a blood stain on the surface of a deck colored with the above mentioned mahogany wood stain they would have many options to choose from. Reds and blacks would get lost in the ambient color unless they substantially altered the intensity (darkness). When picking a color to look at then, a blue would be an obvious choice.

Below shows an overlay of Blueblood™ Tracker and as can be seen, above 650 nm will be substantially different.

Image 46

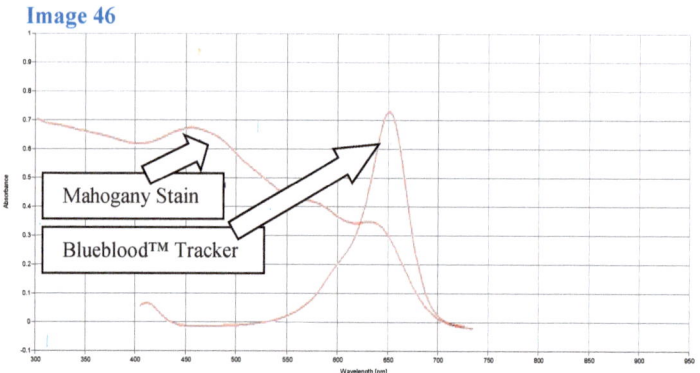

Admittedly the ambient color is not always obvious. A water bottle for example is clear. It would be logical to assume therefore that any color would be strongly visible. But, the "clear" plastic bottle is really slightly yellow. Bluing dyes are added to offset the yellow and pull the position on the blue/yellow color axis back to the origin and give the illusion of clarity. Therefore, a red dye would be the better alternative.

Image 47

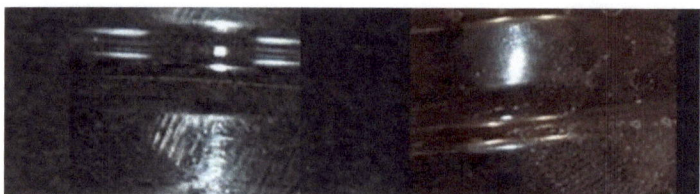

Fingerprint on PET water bottle with and without fuming with AXI Red

One of the advantages of using industrial classifications rather than chemical family to select colorants is that entire industries have been developed around finding different colorants with similar behavior. Factors like solubility and substrate affinity which are critical in industrial application also apply in forensic applications.

Amido Black for example, works because it is soluble in methanol and has dyeing affinity for the dried blood in an acid environment. This is not dissimilar to the dyeing systems typical of other acid dyes. It is reasonable to suppose therefore that other acid dyes might be found which would be compatible with the existing method but which would allow for additional colors.

To support this argument, one need only look at the procedure for the Crowle's Double Stain. (cbdiai.org/reagents) In this procedure Crocien Scarlet 7B (Acid red 71) and Coomassie Brilliant Blue R (Acid Blue 83) are used to dye blood following a procedure similar to the one used with Amido black. Although the procedures are not identical, they are both acidified solutions one in water and one in methanol. Of course Amido Black in a water solution can be used to dye blood. (cbdiai.org/Reagents/crowles)

Similarities can also be found between the use of Leuco Malachite Green and Leuco Crystal Violet. Although both dyes are of different chemical types, they are both water-soluble cationic dyes of the Basic Dyes class. And there are many more dyes in this class. (AATCC 1997)

For all the color options that exist and are in use in forensic laboratories and crime scenes, it is safe to say that only a fraction of the available colors have been identified. And by identifying additional colors, the relative value of existing techniques can be increased. Consider the method presented by Steel and Ball (Steel & Ball 2003) for inversion coloring of the background around fingermarks on plastic tape.

In their method, the full color space was presented because they found three similar dyes in the sublimation class (Disperse Red 60, Disperse Yellow 211 and Disperse Blue 60). All three of these dyes have similar solubility, sublimation and affinity characteristics. As a result, the technique can be used regardless of the ambient background color.

If the same logic can be applied to other techniques then the overall value of those techniques will expand. Consider the Oil Red O process. The process would work better on brown craft and box board if an appropriate solvent blue with oil solubility could be found.

Following the principals listed here, recent work at the University of Illinois at Chicago has identified an oil soluble blue dye similar to Oil Red O that can also evolve fingermarks on porous samples. (youtube.com/watch?v=hRE94Zr6u54)

Even if the analyst does not have the ability or desire to identify new colorants, looking at the existing color space may help direct them to selecting the best technique for the ambient color. And by looking at the color curves of the colorants and ambient background involved the analyst can also select the best optical filters, or the best electronic filters to use on the relevant computer imaging programs.

Whether the forensic scientist is simply looking for evidence on a background, doing research geared to expanding the utility of any specific technique, or simply trying to decide which technique or equipment set to use, considering the specific color of the colorants and background can provide valuable insight.

Beyond just the idea of putting light colors on dark surfaces and dark colors on light surfaces, specific color combination along targeted axes can allow for easier visualization. The forensic scientist should therefore be familiar with color space and the specific color value of the products they use.

Key Points

- **Using industrial classifications alternate colors can be found for existing techniques.**

- **Proper selection of color space will improve visualization.**

References:

American Association of Textile Chemists and Colorists. Technical manual of the American Association of Textile Chemists and Colorists. Vol. 72. The Association, 1997

http://www.cbdiai.org/Reagents

http://www.cbdiai.org/Reagents/crowels.html

https://www.youtube.com/watch?v=hRE94Zr6u54

Steele, Charles, Ball, Mikki., Enhancing Contrast of Fingermarks on Plastic Tape, Journal of Forensic Science, Vol. 48, No. 6, November 2003.

Appendix 1: Descriptions of Images

Image 1: **Color of a Leaf** is a representation of how the perceived color of a leaf is generated. Sunlight containing all available wavelengths of light, represented in the image as blue, yellow and red arrows, are shown striking the leaf. The reflected light is shown to only have some of the available wavelengths, represented as blue and yellow. The combination of the remaining wavelengths combine to form green, the color that the human eye perceives the leaf to be.

Image 2: **Color of Semi-Transparent Objects** is a representation of the color of an object appearing different if one is observing the light that passes through it vs. the light that reflects off of it.

Image 3: **The Forsius Color System** is one of the first color systems emulating the modern view of color. The elementary colors: Red, Yellow, Green and Blue are lines running down the curve of the circle on opposite sides of the center line which goes from white to black. This representation foreshadows the idea of opposing pairs of colors.

Image 4: **Newton Color Wheel** more directly shows the idea of opposed colors as being on opposite sides of the circle. In addition, this system shows blends of the primary colors producing additional shades.

Image 5: **LAB Color Axes** is the foundation for the Hunter and CIE systems. The three dimensional color space described by this image shows that the color axes are perpendicular to each other. Changes along one axis in color space do not affect the coordinate along the other two axes.

Image 6: **Wave Addition** is a representation of the fact that light travels as waves. Like any other wave light waves will constructively and destructively add to form a net wave.

Image 7: **A vs. λ Curves of Three Dyes** the absorbance spectra of three common dyes; Acid Blue 9, Acid Red 30 and Acid Yellow 23.

Image 8: **Monochromatic Acid Brown Dye** shows the absorbance spectra of a common brown dye used in wood stain applications.

Image 9: **Curve Addition of Dyes in Image 7** If the spectra in Image 7 are added, as would be the case if the dyes were blended at a 1:1 ratio, the resulting spectra would be very similar to the spectra presented in Image 8.

Image 10: **Brown Dye Curves** is a theoretical comparison of two random dyes that for the purpose of the discussion are presumed to have a DE > 1 when compared for color.

Image 11: **Anthraquinone Structure** is the defining chemical structure present in all dyes of the Anthraquinone family.

Image 12: **Azo Structure** is the defining chemical structure present in all AZO molecules.

Image 13: **Triphenylmethanes Structure** is the defining chemical structure present in all dyes of this type.

Image 14: **Amido Black Data** presents a standard absorbance curve in water at 40 ppm. Also in the image is the dye's molecular form.

Image 15: **Blueblood™ Tracker Data** presents a standard absorbance curve in water at 40 ppm of the reacted form of the dye, Acid Blue 9. Also in the image is the dye's molecular form of the reacted dye.

Image 16: The top image shows 8 blood foot prints made with an impression stamp on a white plastic surface. The stamp pad touched to blood and pressed onto the surface 8 times. Only the first five impressions were visible

The bottom image shows the same 8 foot prints one minute after being sprayed with Blueblood™ Enhanced. In this image all 8 foot prints are visible.

Image 17: **Crocein Scarlet Data** presents a standard absorbance curve in water at 40 ppm. Also in the image is the dye's molecular form

Image 18: **Fluorescein Data** presents a standard absorbance curve in water at 40 ppm. Also in the image is the dye's molecular form and the fluorescent emission when excited by a Pentax UV light.

Image 19: **Basic Yellow 40 Data** presents a standard absorbance curve in water at 40 ppm. Also in the image is the dye's molecular form and the fluorescent emission when excited by a Pentax UV light.

Image 20: **Leuco Crystal Violet Data** presents a standard absorbance curve in water at 40 ppm of the reacted dye. Also in the image is the reacted dye's molecular form.

Image 21: **Malachite Green Data** presents a standard absorbance curve in water at 40 ppm of the reacted dye. Also in the image is the reacted dye's molecular form.

Image 22: **Rhodamine 6G Data** presents a standard absorbance curve in water at 40 ppm. Also in the image is the dye's molecular form and the fluorescent emission when excited by a Pentax UV light.

Image 23: A single fingerprint was deposited on a piece of glass which was then split in two. The portion on the left was chilled and the portion on the right was left at ambient temperature. Both halves were then fumed cyanoacrylate vapors together in the same chamber at the same time then treated with rhodamine 6G according to the published method. The sample on the left, which had been cooled, accepted more dye and thus developed a stronger color.

Image 24: **Nile Red Data** presents a standard absorbance curve in doixane at 40 ppm. Also in the image is the dye's molecular form.

Image 25: **Solvent Based Color and Fluorescence of Nile Red** presents a general view of the different colors that can be obtained with this dye in different solvents in the top row. The bottom row shows the fluorescent emission when excited at 366 nm. From Left to Right the colors represent the dye in the following solvents: Methanol, Ethanol, Acetonitrile, Dimethylformamide, Acetone, Ethylacetate, Dichloromethane, n-Hexane, Methyl-tert-Butyletherm Cyclohexane, Toluene.

Image 26: **Oil Red O Data** presents a standard absorbance curve in the standard buffered methanol/water used in the Oil Red O process solution, diluted to 40 ppm. Also in the image is the dye's molecular form.

Image 27: **Solvent Yellow 43 Data** presents a standard absorbance curve in acetone at 40 ppm. Also in the image is the dye's molecular form and the fluorescent emission when excited by a Pentax UV light.

Image 28: Is the first fingermarks collected using CN-Yellow.

Image 29: **Fuming Orange Data** is an absorption spectra of the Fuming Orange resin at 40 ppm dissolved in Acetone.

Image 30: **Fuming Orange Fingerprints** Is a single set print processed with the Fuming Orange then illuminated with an ALS at different wavelengths. The images shows are the florescent colors produced at each incident wavelength.

Image 31:   **Fuming Orange vs. Rhodamine** is a single fingermark split in two. The left side of the print was processed with fuming orange. The right side was processed with 2-ethyl cyanoacrylate and rhodamine 6G according to the standard method. As can be seen, the Fuming Orange produces a more consistent image while the rhodamine produces a brighter image.

Image 32: **Axis Inversion Blue Data** presents a standard absorbance curve in ethanol at 40 ppm. Also in the image is the dye's molecular form.

Image 33:   **Axis Inversion Red Data** presents a standard absorbance curve in ethanol at 40 ppm. Also in the image is the dye's molecular form.

Image 34:   **AXI Red on LDPE** shows a fingerprint on an LDPE zip lock bag produced by fuming the bag with Axis Inversion Red. The bag was then placed in an incubator between 35°C to 45°C at 80% humidity for 30 days. At the end of this period, no change in the quality of the fingerprint image was detected.

Image 35:   **Axis Inversion Yellow Data** presents a standard absorbance curve in ethanol at 40 ppm. Also in the image is the dye's molecular form.

Image 36:   **Gel Lifter** shows a fingerprint collected with a BVDA Gel Lifter then treated with Axis Inversion Yellow. The sample was collected by gently pressing a finger to a clean glass surface then applying the gel lifter according to standard methods. The gel lifter was them placed above the fumes of heated AXI Yellow until color stated to develop. The AXI Yellow was able to invade the gel lifter everywhere except where the finger print was blocking the surface. The resulting effect was to shift the color or the gel lifter 2 shade units to the yellow everywhere except under the print. The picture shown was taken with an oblique angle to allow the normal contrast of the gel lifter to be visible as well.

Image 37   The molecular form if ABTS.

Image 38:   The molecular form of Silver Nitrate.

Image 39:   The molecular form of Diaminobenzidine.

Image 40:   The molecular form of Ninhydrin.

Image 41:   The molecular form of Ninhydrin after it reacts.

Image 42:   The molecular form of DFO.

Image 43:   The molecular form of Phenolphthalein.

Image 44: Presents a standard absorbance curve of carbon black locked in a clear styrene matrix.

Image 45: Presents a standard absorbance curve of a mahogany wood stain at 40 ppm in water.

Image 46: Presents the wood stain from image 45 overlain with a Blueblood™ Tracker spectra to demonstrate which wavelength would be best to visualize blood deposited on a mahogany surfaces which had then been treated with Blueblood™ Tracker or Blueblood™ Enhanced.

Image 47: Shows a fingerprint on a PET water bottle which was fumed with cyanoacrylate then treated with Axis Inversion Red. The image on the left is the water bottle after cyanoacrylate fuming alone. The image on the right is part of the same print after fuming with Axis Inversion red.

Appendix 2: Glossary

AATCC

American Association of Textile Chemists and Colorists

A not-for-profit association providing worldwide: test method development, quality control materials, and professional networking for the textile industry.

ABSORPTION SPECTROSCOPY

Techniques measuring a sample's absorption of photons at specific wavelengths or frequencies.

ACID DYE

Water soluble direct colorant, frequently a sodium or ammonium salt of a sulfuric, carboxylic or phenolic organic acid. Acid dyes typically possesses good affinity for amphoteric fibers. Dyeing methods use ionic bonding with fiber cationic sites to fix the color to the substrate. Acids are added to dyeing baths to increase the number of bonding locations. Acid dyes are used in many applications including cleaners, cosmetics and food colorants.

ALBUMIN

Family of water soluble proteins found in blood.

AMBIENT LIGHT

Illumination present in an area without bringing in additional light sources.

ANIONIC

Containing a negatively charged ion.

AXIS INVERSION DYES

Disperse dyes designed to color polymer surfaces around a fingermark.

BAND-PASS FILTERS

Colored lenses that block certain wavelengths of light. They are usually produced to only allow narrow ranges of light to pass through.

BASIC DYE

(Also known as cationic dyes) This class of synthetic dyes includes bases which when made to be soluble in water form a colored cationic salt. Basic dyes bind to anionic sites on the surface of the substrate and produce bright shades with high tinctorial values, on textile.

BEER-LAMBERT LAW

An equation for relating the absorption to the quantity of matter through which the photons are passed.

BLOOD SPATTER

Blood residue left at a crime scene.

CATIONIC

Containing a positively charged ion.

CHEMILUMINESCENCE

The generation of visible light photons as a result of chemical reactions.

CIELAB

A color measurement system separating color space into three perpendicular axes. L is a measure of dark (L=0) to light (L=100). "a" is the red/green axis. −a is green +a is red. "b" is the blue/yellow axis. −b is blue +b is yellow.

CMYK

Process color system combining Cyan, Magenta, Yellow and Black base colors to form the dull range of shades. These system are common for printing applications.

CN-YELLOW

A precolored cyanoacrylate resin used for fingerprint fuming applications to provide one step processing.

COLOR

The Perception of visible light photons.

COLOR ACUITY

A measure of how well a person perceives specific colors.

COLORANT

Substances used to impart color to something else, i.e., Pigments, Dyes and Stains.

COLORIMETERS

Devices used to measure the strength of a color at a specific wavelength.

COMISSION INTERNATIONAL DE L'ECLARIGE (CIE)

The International Commission on Illumination - also known as the CIE from its French title, the Commission Internationale de l'Eclairage - is devoted to worldwide cooperation and the exchange of information on all matters relating to the science and art of light and lighting, colour and vision, photobiology and image technology.

CONES

Component of the eye that perceives color

CYANOACRYLATE

A polymer used primarily as an adhesive but which has been found to sublime and have affinity for biologic residue where it forms a hard surface in the shape of the residue. It is used to develop latent fingermarks.

DE

The relative measure of the color difference between two points in color space. (See CIELab)

DIODE ARRAYS

Detectors in spectrophotometers that use a series of photo diodes to measure the amount of light striking the detector.

DISPERSE DYE

Special class of solvent dyes that are frequently uses in water dispersions to dye polyester and eyeglasses.

DNA TESTING (FORENSIC)

The forensic science dedicated to identifying persons or sources of biological material from DNA residue

DR. ALEC JEFFREYS

Professor Sir Alec Jeffreys is a biochemist and geneticist at the University of Leicester where he currently holds the positions of Professor of Genetics and Royal Society Wolfson Research Professor.

Professor Jeffreys' research at Leicester focuses on exploring human DNA diversity and the mutation processes that create this diversity. He was one of the first to discover inherited variation in human DNA, then went on to invent DNA fingerprinting, showing how it can be used to resolve issues of identity and kinship and created the field of forensic DNA. The subsequent impact of DNA on solving paternity and immigration cases, catching criminals and freeing the innocent has been extraordinary, directly affecting the lives of millions of people worldwide.

DYE

A direct colorant that imparts color when dissolved into the substance it is coloring.

EXTINCTION COEFFICIENT

The ratio between the amount of light absorbed by a sample and its molar mass.

FINGERMARK

A copy of the ridge impressions of a fingertip left on a surface of contact.

FINGERPRINT

A visible copy of the ridge impressions of a fingertip left on a surface of contact.

FLUORESCENT

The light emitted from a substance after it has been exposed to ambient radiation.

FORENSIC SCIENTIST

A scientists who's discipline interacts with the courts.

FORMS ACUITY

The ability of a person to discern unique shapes.

FUMING ORANGE

A precolored cyanoacrylate resin used for fingerprint fuming applications to provide one step processing.

GEL LIFTERS

A low tack adhesive gelatin layered film used to recovering samples like footprints and fingerprints one layer at a time.

IMMUNOGLOBULINS

The antibodies produced in cells

INNOCENCE PROJECT

A national litigation and public policy organization working to exonerate wrongfully convicted individuals.

OXIDIZER

Component that accepts an electron promoting a chemical reaction.

PHOTOMULTIPLIER DETECTOR

Device for increasing the charge caused by photon collusion so that it can be recorded.

PIGMENT

Insoluble colorant

POLYCYANOACRYLATE

The polymerized from of cyanoacrylate

REFLECTION

Uniform redirection of photons that have interacted with a sample.

SCATTERING

Non-uniform redirection of photons that have interacted with a sample.

SPECTROPHOTOMETERS

Device for measuring the amount of light absorbed by a sample.

SPECTRUM ANALYZERS

Device for measuring the composition of light

SUBLIME

Transitioning from a solid state to a vapor state without liquefying

THE SOCIETY OF DYERS AND COLOURISTS (SDC)

Established in 1884, the SDC is the leading independent, educational charity dedicated to advancing the science and technology of colour.

TINCTORIAL

Relating to dyeing or staining; imparting color